COMPLEXITY OF
COMPUTER COMPUTATIONS

THE IBM RESEARCH SYMPOSIA SERIES

1971: Computational Methods in Band Theory
Editors: P. M. Marcus, J. F. Janak, and A. R. Williams

1972: Computational Solid State Physics
Editors: F. Herman, N. W. Dalton, and T. R. Koehler

1972: Sparse Matrices and Their Applications
Editors: D. J. Rose and R. A. Willoughby

1972: Complexity of Computer Computations
Editors: R. E. Miller and J. W. Thatcher
Associate Editor: J. D. Bohlinger

COMPLEXITY OF COMPUTER COMPUTATIONS

Proceedings of a symposium on the Complexity of Computer Computations, held March 20-22, 1972, at the IBM Thomas J. Watson Research Center, Yorktown Heights, New York, and sponsored by the Office of Naval Research, Mathematics Program, IBM World Trade Corporation, and the IBM Research Mathematical Sciences Department

Editors

Raymond E. Miller • James W. Thatcher

Associate Editor

Jean D. Bohlinger

Mathematical Sciences Department
IBM Thomas J. Watson Research Center
Yorktown Heights, New York

PLENUM PRESS • NEW YORK - LONDON • 1972

Library of Congress Catalog Card Number 72-85836
ISBN 0-306-30707-3

A Division of Plenum Publishing Corporation
227 West 17th Street, New York, N.Y. 10011

United Kingdom edition published by Plenum Press, London
A division of Plenum Publishing Company, Ltd.
Davis House (4th Floor), 8 Scrubs Lane, Harlesden, London, NW10 6SE, England

Printed in the United States of America

PREFACE

The Symposium on the Complexity of Computer Computations was held at the IBM Thomas J. Watson Research Center in Yorktown Heights, New York, March 20-22, 1972. These Proceedings contain all papers presented at the Symposium together with a transcript of the concluding panel discussion and a comprehensive bibliography of the field.

The Symposium dealt with complexity studies closely related to how computations are actually performed on computers. Although this area of study has not yet found an appropriate or generally accepted name, the area is recognizable by the significant commonality in problems, approaches, and motivations. The area can be described and delineated by examples such as the following.

(1) Determining lower bounds on the number of operations or steps required for computational solutions of specific problems such as matrix and polynomial calculations, sorting and other combinatorial problems, iterative computations, solving equations, and computer resource allocation.

(2) Developing improved algorithms for the solution of such problems which provide good upper bounds on the number of required operations, along with experimental and

theoretical evidence concerning the efficiency and numerical accuracy of those algorithms.

(3) Studying the effects on the efficiency of computation brought about by variations in sequencing and the introduction of parallelism.

(4) Studying the relative complexity of classes of problems with respect to lower bounds on computation time. In this effort, specific problems are classified as having equivalent difficulty of computation; for example, those problems which can be solved in a number of operations which is a polynomial function of the size of the input.

The Symposium succeeded in bringing together a large number of people interested in this important field. Among the 225 registrants, there were many who are actively involved in this and related research together with those generally interested in more efficient utilization of computers. We expect that the stimulation arising from discussions between attendees will have a significant effect on further work in the field.

We hope that through these Proceedings the Symposium will also provide a broad picture of current research efforts in the field for a much larger audience. The collection of papers, which is the first of its kind, should serve as a good research reference and, combined with the comprehensive bibliography and stimulating panel discussion, will hopefully be useful for courses and seminars on the topic.

We are indebted to the Mathematical Sciences Department of IBM Research, the Mathematics Program of the Office of Naval Research, and the IBM World Trade Corporation for their support in cosponsoring the Symposium.

We wish to express our thanks to the authors for making their papers available while complying with strict deadlines and formats to aid in the timely appearance of the book; to Michael Rabin and Shmuel Winograd who participated in the initial planning of the Symposium; to Edward M. Reingold for supplying us with his extensive bibliography which, in combination with the other authors' references, comprises the bibliography for these

Proceedings; and to Jean Bohlinger for her extensive help in organizing the Symposium as well as serving as Associate Editor of this volume.

R. E. Miller
J. W. Thatcher

May 1972

CONTENTS

EVALUATION OF RATIONAL FUNCTIONS

Volker Strassen

Universität Zürich

Summary

In the first part of this paper the complexity (with respect to multiplication and division) of a general continued fraction and of arbitrary quadratic forms is determined with the help of Pan's method. In the second part some results on avoiding division and on the multiplication of matrices from an algebraic group are reported.

Pan's Method

Let k be an infinite field, $X_1,\dots,X_s$, $x_1,\dots,x_n$ indeterminates over k. In the sequel we often use X,Y as names for X_1,X_2 and a_i,b_j,a_{ij} as names for pairwise different x_ν. We consider $A=k[X_1,\dots,X_s,x_1,\dots,x_n]$ as k-ring over $k[X_1,\dots,X_s]\cup\{x_1,\dots,x_n\}$, i.e. we allow the binary operations +,-,*, the unary operations "multiplication with an element of k" and the elements of $k[X_1,\dots,X_s]\cup\{x_1,\dots,x_n\}$ as 0-ary operations (constants).

We assume that it takes one time-unit to perform the operation $*$, and no time to perform any other operation (including the multiplication by scalars). We have the computational length function L, which assigns to each finite subset F of A the optimal time to compute F serielly (see Winograd (1970A), Strassen (1972/73A)).

The main reason for considering only the operation $*$ as time-consuming is mathematical convenience. Of course any lower bound for L in our situation remains a lower bound if the other operations are also taken into account. In some cases the special treatment of $*$ is realistic (even its distinction from the multiplication by scalars), e.g. if the indeterminates X_σ, x_ν take large (commuting) matrices as values.

The essence of the formal notion of a computation in the context of numerical analysis is due to Ostrowski (1954A). As is well known, Ostrowski conjectured in his paper that Horner's rule for computing a general polynomial is optimal. This conjecture has been proved by Pan (1966A):

<u>1 Theorem</u> $L(\sum_{i=0}^{q} a_i X^i) = q.$

Pan's theorem immediately implies several related results:

<u>2 Corollary</u> $L(\{\sum_{j=0}^{q} a_{ij} X^j : 1 \le i \le p\}) = pq.$

<u>Proof</u> If $L(\{\sum_{j=0}^{q} a_{ij} X^j : 1 \le i \le p\}) < pq$ then $L(\sum_{i=1}^{p} X^{(q+1)(i-1)} \sum_{j=0}^{q} a_{ij} X^j) < pq+p-1$ since we work over $k(X)$. The left side of this inequality is the computational

length of a general polynomial of degree qp+p-1. Therefore the inequality contradicts Pan.

3 Corollary $L(\sum_{o\leq i,j\leq q} a_{ij}X^iY^j) = (q+1)^2-1$.

Proof Substitute X^{q+1} for Y. Since this substitution is a homomorphism of k-rings mapping $k[X_1,\ldots,X_s]\cup\{x_1,\ldots,x_n\}$ into itself we get $L(\sum a_{ij}X^iY^j)\geq$ $\geq L(\sum a_{ij}X^{i+(q+1)j}) = (q+1)^2-1$ by Pan.

4 Corollary $L(\{\sum_{1\leq j\leq q} a_{ij}X_j : 1\leq i\leq p\}) = pq$

Proof Substitute X^j for X_j and use corollary 2.

Actually, Pan also allows division, which is counted like multiplication. The substitutions in the last two proofs then have to be modified somewhat. The above corollaries have been obtained by Winograd (1970A) from a more general theorem than Pan's. Let $k\subset K$ be infinite fields, $x_1,\ldots,x_n$ indeterminates over K. Consider $K(\underline{x})=K(x_1,\ldots,x_n)$ as k-field over $K\cup\{x_1,\ldots,x_n\}$ (i.e. with +,-,*,/, scalar multiplications and the elements of $K\cup\{x_1,\ldots,x_n\}$ as operations). * and / are counted. Winograd's theorem applies to any finite set of linear forms in $K(\underline{x})$. Below, we will give a result that works for arbitrary rational functions. For the proof we use Pan's method, which consists in eliminating the (*/)-operations inductively by suitable substitutions of the variables.

A subset of K^n is called (Zariski-) dense, if it is not contained in the set of zeros of a nonvanishing polynomial.

5 Theorem Let $F=\{f_1,\ldots,f_r\}\subset K(\underline{x})$, $L(F)\leq m\leq n$. Then there

is a dense subset G of K^n with the following property: For every $\underline{v}\in G$ there are an $n\times(n-m)$-matrix Γ over k, an $r\times(n-m)$-matrix Λ over k, and $\underline{w}\in K^r$ such that Γ has maximal rank, $\underline{f}(\Gamma\underline{t}+\underline{v})=(f_1(\Gamma\underline{t}+\underline{v}),\ldots,f_r(\Gamma\underline{t}+\underline{v}))$ is defined and

$$\underline{f}(\Gamma\underline{t}+\underline{v}) = \Lambda\underline{t}+\underline{w}$$

($t_1,\ldots,t_{n-m}$ are indeterminates over K and $\underline{t}=(t_1,\ldots,t_{n-m})$).

Remarks The case r=1, n-m=1 will be most frequently used. Here we write γ_i,t,λ,w for $\gamma_{i1},t_1,\gamma_{11},w_1$. As is easily seen the theorem remains true if everywhere the total number n of indeterminates is replaced by the number of indeterminates (say q) actually occurring in the elements of F.

Proof View elements $f\in K(x_1,\ldots,x_n)$ as partial functions $K^n \to K$, whose domain Def(f) consists of all $\underline{v}\in K^n$ at which the (reduced) denominator of f does not vanish. Call an affine linear function $\gamma_1x_1+\ldots+\gamma_nx_n+v$ a good function, if $\gamma_i\in k$ for all i, $v\in K$. Call an affine subspace L of K^n a good subspace, if it is the set of common zeros of a set of good functions. This is equivalent to saying that there are good functions $\ell_1,\ldots,\ell_n$ on $K^{\dim L}$ such that L is the image of the affine linear map $(\ell_1,\ldots,\ell_n) : K^{\dim L} \to K^n$. In order to prove the theorem, it will suffice to prove the following geometric version: Given a finite set $F\subset K(x_1,\ldots,x_n)$, then for any m with $L(F)\le m\le n$ and any nonzero $h\in K[x_1,\ldots,x_n]$ there is a good (n-m)-dimensional subspace L of K^n hitting $\{h\neq 0\}$ such that for any $f_\rho\in F$ $\mathrm{Def}(f_\rho)\cap L\neq\emptyset$ and there is a good function $\tilde{f}_\rho$ on K^n which coincides with f_ρ on $\mathrm{Def}(f_\rho)\cap L$. We prove this by L-induction (see

Strassen (1972/73A)):The induction step being trivial for the constant and linear operations, we restrict ourselves to / (* can be treated similarly). Let the above statement be true for the set $\{f_1,\ldots,f_{r+2}\}$ and assume $f_{r+2}\neq 0$, $L(\{f_1,\ldots,f_r,f_{r+1}/f_{r+2}\})=L(\{f_1,\ldots,f_{r+2}\})+1$. We have to show that the statement also holds for $\{f_1,\ldots,f_r,f_{r+1}/f_{r+2}\}$. Let $L(\{f_1,\ldots,f_r,f_{r+1}/f_{r+2}\})\leq m\leq n$ and let $h\in K[x_1,\ldots,x_n]$ be nonzero, $U=\{h\neq 0\}$. By induction hypothesis there is a good $(n-m+1)$-dimensional subspace L' of K^n such that $U\cap L'\neq\emptyset$, $\mathrm{Def}(f_\rho)\cap L'\neq\emptyset$ and $f_\rho=\tilde{f}_\rho$ on $\mathrm{Def}(f_\rho)\cap L'$ with good $\tilde{f}_\rho$ $(1\leq\rho\leq r+2)$. If $\tilde{f}_{r+1}$ and $\tilde{f}_{r+2}$ are both constant, take any good $(n-m)$-dimensional subspace L of L' hitting U and $\mathrm{Def}(f_\rho)$. If say $\tilde{f}_{r+2}$ is not constant, then take $L=\{f_{r+2}=\alpha\}\cap L'$, where $\alpha\in k-\{0\}$ is chosen so that $\mathrm{Def}(f_\rho)\cap L\neq\emptyset$ and $U\cap L\neq\emptyset$.

To get Pan's result from theorem 5, let $K=k(X)$. Replace $\sum_{i=o}^{q} a_i X^i$ by $\sum_{i=1}^{q} a_i X^i$ and assume

$$L\Big(\sum_{i=1}^{q} a_i X^i\Big)\leq q-1.$$

Then we have

$$\Big(\sum_{i=1}^{q}\gamma_i X^i\Big)t+\sum_{i=1}^{q} v_i X^i = \lambda t+w,$$

thus

$$\sum_{i=1}^{q}\gamma_i X^i = \lambda\in k,$$

contradicting the fact that not all γ_i are 0.

In a similar way one deduces corollaries 2-4 and the following

<u>6 Corollary</u> Let $K=k(X_1,\ldots,X_q)$ (recall that the elements of K are treated as constants). Then

$$L\left(\sum_{1\leq i\leq j\leq q} a_{ij}X_iX_j\right) = \frac{1}{2}q(q+1).$$

From now on we will always have K=k.

<u>7 Corollary</u> Let Char k$\neq$2, $f = \sum_{1\leq i,j\leq n} \tau_{ij}x_ix_j$ with $\tau_{ij}=\tau_{ji}\in k$. Then $L(f) = n-p$, where p is the dimension of a maximal nullspace of f (by Witt's theorem, p does not depend on the choice of such a nullspace).

<u>Proof</u> W.l.o.g. f nonsingular. Put m=L(f). As is well known

$$f \sim x_1x_2+\ldots+x_{2p-1}x_{2p} + \alpha_{2p+1}x_{2p+1}^2+\ldots+\alpha_n x_n^2$$

with $\alpha_i\in k$ ($\sim$ denotes equivalence of quadratic forms). This yields $m\leq n-p$. By theorem 5 we have

$$\sum_{i,j\leq n} \tau_{ij}\left(\sum_{\nu=1}^{n-m}\gamma_{i\nu}t_\nu+v_i\right)\left(\sum_{\mu=1}^{n-m}\gamma_{j\mu}t_\mu+v_j\right) = \sum_{\nu=1}^{n-m}\lambda_\nu t_\nu+w.$$

Comparing coefficients we get

$$\sum_{i,j}\tau_{ij}\gamma_{i\nu}\gamma_{j\mu} = 0$$

for all ν,μ. Thus the (n-m)-dimensional linear subspace of k^n generated by the vectors $(\gamma_{1\nu},\ldots,\gamma_{n\nu})$ is a nullspace for f. This implies $p\geq n-m$.

Corollary 7 has been proved in Strassen (1973A) by a different method. Special cases: if $k=\mathbb{C}$ and f has rank q, then

(1) $$L(f) = \left\lceil \frac{q}{2}\right\rceil ,$$

if $k=\mathbb{R}$ and f has signature (π,ν), then

(2) $$L(f) = \max\{\pi,\nu\}.$$

These two facts also appear in Vari (1972A).

<u>8 Corollary</u> Let f_n be the continued fraction formed

with $x_n,\dots,x_1$ (i.e. $f_1=\frac{1}{x_1}$, $f_{j+1}=1/(x_{j+1}+f_j)$). Then

$$L(f_n) = n.$$

Proof If $L(f_n)\le n-1$, then we have

(3) $$f_n(\gamma_1 t+v_1,\dots,\gamma_n t+v_n) = \lambda t+w.$$

We use the following simple fact about continued fractions (see Davenport (1970A)): A representation of f_n as reduced quotient of polynomials is

(4) $$f_n = \frac{[x_{n-1},\dots,x_1]}{[x_n,\dots,x_1]},$$

where $[x_n,\dots,x_1]$ is the sum of $x_n\dots x_1$ and all subproducts of $x_n\dots x_1$ which can be obtained by deleting disjoint pairs of the form $x_i x_{i-1}$. Put

$$I_n(t) = [\gamma_n t+v_n,\dots,\gamma_1 t+v_1].$$

Then (3) becomes

(5) $$I_{n-1}(t) = (\lambda t+w)I_n(t).$$

By suitable choice of $\underline{v}$ we have $I_{n-1}(t) \neq 0$.

Let $\gamma_{i_1},\dots,\gamma_{i_p}$ be the nonvanishing elements in the sequence $\gamma_1,\dots,\gamma_n$. Then $I_n(t)$ has degree $\le p$ and p'th coefficient

$$\gamma_{i_1}\dots\gamma_{i_p}\, g(\underline{v}),$$

where $g(\underline{x})$ is a nonzero polynomial of degree $n-p<n$. By suitable choice of $\underline{v}$ we have

$$g(\underline{v})\neq 0,$$

thus I_n has degree p. Since I_{n-1} has degree $\le p$, and $\le p-1$ if $\gamma_n\neq 0$, we conclude

$$\gamma_n = \lambda \doteq 0.$$

The p'th coefficient of I_{n-1} is

$$\gamma_{i_1}\dots\gamma_{i_p}\cdot h(\underline{v}),$$

where $h(\underline{x})$ is a polynomial. Comparing 0'th coefficients in (5) we get

(6) $$[v_{n-1},\ldots,v_1] = w[v_n,\ldots,v_1],$$

comparing p'th coefficients

(7) $$h(\underline{v}) = wg(\underline{v}).$$

Now

$$f_n \neq \frac{h(\underline{x})}{g(\underline{x})},$$

since degree(g)<n and the representation (4) is reduced. Thus by suitable choice of $\underline{v}$

$$[v_{n-1},\ldots,v_1]g(\underline{v}) \neq [v_n,\ldots,v_1]h(\underline{v}),$$

contradicting (6), (7).

<u>9 Corollary</u> $L(\sum_{i=1}^{n} \frac{1}{x_i}) = n.$

<u>Proof</u> similar to the last proof.

Let $\sigma_1,\ldots,\sigma_n$ be the elementary symmetric functions of $x_1,\ldots,x_n$. Thus

$$(t-x_1)\ldots(t-x_n) = t^n-\sigma_1(\underline{x})t^{n-1}+\ldots+(-1)^n\sigma_n(\underline{x}),$$

where t is an additional indeterminate. Clearly $L(\sigma_1)=0$, and from corollary 7

(8) $$L(\sigma_2) = \begin{cases} \lceil \frac{n}{2} \rceil & \text{if } k = \mathbb{C} \\ n-1 & \text{if } k = \mathbb{R}, \end{cases}$$

since σ_2 has signature (1,n-1). Moreover

(9) $$L(\sigma_n) = n-1,$$

for otherwise $\prod_{i=1}^{n}(\gamma_{i1}t_1+\gamma_{i2}t_2+v_i)=\lambda_1t_1+\lambda_2t_2+w$, which is impossible for a matrix Γ of rank 2 if we choose all $v_i \neq 0$. We remark also that

(10) $L(\{\sigma_1,\ldots,\sigma_n\}) \leq n(^2\lg n)^{\cdot}$,

as may be shown by going from n to 2n using interpolation.

Further Results

The following results will appear in Strassen (1973A). If F is a set of polynomials in $k(\underline{x})=k(x_1,\ldots,x_n)$, then besides $L(F)=L_{k(\underline{x})}(F)$ one also has $L_{k[\underline{x}]}(F)$, i.e. the computational length of F in the k-ring $k[\underline{x}]$ over $k\cup\{x_1,\ldots,x_n\}$. Obviously $L(F)_{k(\underline{x})} \leq L_{k[\underline{x}]}(F)$.

<u>10 Theorem</u> Let F be a finite set of polynomials of degree $\leq d$. Then

$$L_{k[\underline{x}]}(F) \leq (2d-3)(2\lceil ^2\lg d\rceil-1)L_{k(\underline{x})}(F).$$

In particular, division does not help for the computation of a set of quadratic forms (according to Winograd, P. Ungar (private communication) has also proved this fact in the case, where all quadratic forms are bilinear with respect to a given partition of the set of indeterminates). An example is matrix multiplication. Theorem 10 is proved by developing elements of $k(\underline{x})$ into formal power series at a suitable point and using the formula $(1-f)^{-1} = \sum_{r\geq o} f^r$, f being a power series without constant term. As is immediate and well known, computations not using division of a set of quadratic forms may be reduced to having a very simple structure, namely that of taking linear combinations of products of pairs of linear forms. For this reason, the computational length of a set of quadratic forms can essentially be expressed by a purely algebraic quantity, the rank of the coefficient tensor of the set of quadratic forms (we assume that $\mathrm{Char}\,k \neq 2$ and that the forms are given by symmetric coefficient matrices). The rank of a tensor $\tau \in U \otimes V \otimes W$, where U, V, W are finite dimensional vector spaces, is defined by

$$\text{(11)} \qquad \operatorname{rank}(\tau) = \min\{N : \exists u_1,\dots,u_N \in U \ \exists v_1,\dots,v_N \in V \ \exists w_1,\dots,w_q \in W \quad \tau = \sum_{q=1}^{N} u_q \otimes v_q \otimes w_q\},$$

generalizing one of the possible definitions of the rank of a matrix. The rank is also useful in connection with the problem of multiplying special matrices:

<u>11 Theorem</u> The computational length of the multiplication of two matrices from an irreducible linear algebraic group is at least $\frac{1}{4}$ times the rank of the structural tensor of the Lie ring of the group.

Essentially a special case is the following fact: The computational length of the multiplication of two orthogonal matrices (division allowed) is not less than $\frac{1}{2}$ times the computational length for multiplying two skewsymmetric matrices of the same size (without division). This in turn is easily estimated by the computational length for multiplying two general matrices of half the size.

SOLVING LINEAR EQUATIONS BY MEANS OF SCALAR PRODUCTS

Michael O. Rabin

Institute of Mathematics

The Hebrew University of Jerusalem

1. INTRODUCTION

We shall state and solve a problem proposed by R. Brent and P. Wolfe concerning the solution of a system of linear equations by algorithms where the only operation permitted on the coefficient vectors $a_1, a_2, \ldots,$ is the formation of scalar products $a_i \cdot v$. Each such calculation is called a test. The objective is to solve the equations using a minimum number of tests.

Let R denote the field of real numbers, and let R^n denote Euclidean n-space. Vectors in R^n will be denoted by $a = (a_1, \ldots, a_n)$. The scalar product $\sum a_i b_i$ will be denoted by $a \cdot b$. Let $a_i = (a_{i,1}, \ldots, a_{i,n}) \in R^n$, $1 \le i \le n-1$, be $n-1$ linearly independent vectors and let $x = (x_1, \ldots, x_n)$. Denoting $L_i(x) = a_i \cdot x$, $1 \le i \le n-1$, consider the system

$$(1) \qquad L_1(x) = 0,\ L_2(x) = 0, \ldots, L_{n-1}(x) = 0,$$

of $n-1$ homogeneous equations. These equations have, up to a factor of proportionality, a unique (non-trivial) solution.

Instead of using the individual coefficients a_{ij} of the equations, we restrict ourselves to tests of the following form. A test consists of taking a numerical vector $v \in R^n$ and finding, for some $1 \le i \le n$, the value of $a_i \cdot v = \alpha$.

Clearly, by performing $n(n-1)$ tests we can find all the coefficients and then solve (1). In section 2 we shall see that $n(n+1)/2-1$ tests are sufficient for solving (1). Brent and Wolfe raised the question whether this number of tests is the smallest number necessary in an algorithm for solving (1). We give a precise definition for the notion of an algorithm using tests for solving (1), and prove that $n(n+1)/2-1$ tests are always needed.

The same result holds if (1) is a system of $n-1$ non-homogeneous linear equations

$$(2) \qquad \ell_i(x) = a_{i,1}x_1+\dots+a_{i,n}x_{n-1} + a_{i,n} = 0, \ 1 \le i \le n-1,$$

in $n-1$ unknowns. In fact, a solution of (2) is obtained from a solution $(x_1,\dots,x_n)$ of (1) by dividing by x_n. The result $L_i(v_1,\dots,v_n)$ of a test on $L_i(x)$ can be obtained as $v_n\ell_i(v_1/v_n,\dots,v_{n-1}/v_n)$ from a test on $\ell_i(x_1,\dots,x_{n-1})$. Thus the treatment of (2) can be derived from the results for the homogeneous case (1) and we restrict ourselves to the latter case.

The problem arose with Brent and Wolfe in the following form. They considered a system of equations

$$(3) \qquad P_i(x_1,\dots,x_{n-1}) = 0, \ 1 \le i \le n-1.$$

The function P_i may be written as $P_i = Q_i + \ell_i(x_1,\dots,x_{n-1})$, where $\ell_i(x)$ is the linear part of P_i. It is known that we are dealing with functions for which

$$|Q_i(x)| << |\ell_i(x)| \ , \ 1 \le i \le n-1.$$

Thus a solution $(x_1,\dots,x_{n-1})$ of (2) is a good approximation to a solution of (3). Now, the linear forms $\ell_i(x)$ are not given, but we can test them by computing $P_i(v)$ and assuming $\ell_i(v) = P_i(v)$. Thus we are in a situation of having to solve linear equations by testing the linear forms. Since the functions $P_i(x_1,\dots,x_{n-1})$, $1 \le i \le n-1$, were actually given as solutions of optimization problems depending on the parameters $x_1,\dots,x_{n-1}$, each test (evaluation) $P_i(v)$ is a costly computation. Hence our interest in solving the equations using a minimal number of tests.

During the Symposium it was pointed out to me by G.H. Golub that several known procedures for solving linear equations actually proceed through evaluation of scalar products with the coefficient vectors, i.e. through tests. Namely, the coefficients $a_i \in R^n$ of $L_i(x)$ are not directly stored in memory, but are generated by programs P_i, $1 \le i \le n-1$. The basic step is the formation of products $a_i \cdot v$ with vectors $v \in R^n$ generated during the computation. Thus our results yield information about the minimum computational effort involved in methods of this type.

There may also be implications concerning the memory-space required by algorithms for the solution of a system of linear equations. We did not have time to fully explore the consequences of these observations.

2. A METHOD OF SOLUTION

We shall give a method for solving the equations (1) by means of tests. The method resembels Gaussian elimination and is as simple.

Let $a, v_1, \ldots, v_k \in R^n$. Define

$$N(a) = \{x \mid x \in R^n, \quad a \cdot x = 0\}$$

$$\langle v_1, \ldots, v_k \rangle = \underline{\text{subspace spanned by}} \quad v_1, \ldots, v_k.$$

$N(a)$ is the <u>null space</u> of $\underline{a}$.

LEMMA 1. Let $a \neq 0$, and let $v_1, \ldots, v_k$ be linearly independent. Put $\alpha_i = a \cdot v_i$, $1 \leq i \leq k$, and assume $\alpha_1 = \alpha_2 = \ldots = \alpha_j = 0$, $\alpha_{j+1} \neq 0, \ldots,$ $\alpha_k \neq 0$ ($j = 0$ or $j = k$, is possible); put $\beta_m = \alpha_m / \alpha_{m+1}$, $j + 1 \leq m \leq k-1$. The vectors $v_1, \ldots, v_j$, $v_{j+1} - \beta_{j+1} v_{j+2}, \ldots, v_{k-1} - \beta_{k-1} v_k$, are linearly independent and are a basis for the space $N(a) \cap \langle v_1, \ldots, v_k \rangle$.

The proof is obvious. Note that the number of basis vectors is $k-1$, except in the case that $a \cdot v_i = 0$, $1 \leq i \leq k$.

The algorithm proceeds as follows. Let $a_1, \ldots, a_{n-1} \in R^n$ be the coefficient vectors and let $e_1, \ldots, e_n \in R^n$ be some fixed basis of R^n. Form the n products $a_1 \cdot e_j$, $1 \leq j \leq n$. From these compute a basis $v_{1,1}, \ldots, v_{1,n-1}$, for $N(a_1)$ as in Lemma 1. Form the $n-1$ products $a_2 \cdot v_{1,j}$, $1 \leq j \leq n-1$. Compute a basis $v_{2,1}, \ldots, v_{2,n-2}$ for $N(a_2) \cap \langle v_{1,1}, \ldots, v_{1,n-1} \rangle = N(a_1) \cap N(a_2)$. The number of vectors does diminish by 1 at each step because the linear independence of $a_1, \ldots, a_{n-1}$ implies that $\dim (N(a_1) \cap \ldots \cap N(a_k)) = n-k$, $1 \leq k \leq n-1$. The total number of tests is $n+n-1+\ldots+2 = n(n+1)/2-1$.

We see that in the above algorithm, one equation was tested n times, one equation was tested $n-1$ times, etc. Our theorems will show that this situation will hold for every solution by means of tests of linear equations.

3. ESSENTIAL SETS

The algorithm given in Section 2 will solve the equations (1) by $n(n+1)/_2-1$ tests whenever the coefficient vectors $a_1,\ldots,a_{n-1}$ are linearly independent. Consider $(a_1,\ldots,a_{n-1}) = (a_{1,1},\ldots,a_{1,n},\ldots,a_{n-1,1},\ldots,a_{n-1,n}) = a$, as a vector in R^m, $m = n(n-1)$. If the determinant $Q(a) = |a_{ij}|_{1\leq i,j\leq n-1}$ does not vanish, then the vectors are linearly independent. Thus the set of $a \in R^m$ for which the algorithm works includes all the $a \in R^m$ for which $Q(a) \neq 0$.

In fact, for many algorithms involving parameters $x = (x_1,\ldots,x_m) \in R^m$, there exists a non-trivial polynomial $Q(x)$ such that the algorithm is valid whenever $Q(x) \neq 0$. For example, $Q(x)$ may be the product of all expressions appearing in denominators of quotients, so that if $Q(x) = 0$ then the algorithm may fail, but if $Q(x) \neq 0$ then it works.

These examples lead to the following

DEFINITION 1. A set $H \subseteq R^m$ is <u>essential</u> if for some polynomial $Q(x_1,\ldots,x_m)$ such that $Q(x) \not\equiv 0$ we have

$$Q(a) \neq 0 \ \underline{\text{implies}}\ a \in H.$$

We say that a property P (such as the validity of an algorithm) holds <u>essentially</u> in R^m if the set of $a \in R^m$ for which P is true is an essential set.

A set $S \subseteq R^m$ is called <u>inessential</u> if $R^m - S$ is an essential set.

Essential sets have the following properties which we state, leaving the proof to the reader.

LEMMA 2. If $B(A_1,\ldots,A_k)$ is a boolean expression in the set-variables $A_1,\ldots,A_k$, and $H_i \subseteq R^m$, $1 \leq i \leq k$, are essential sets, then $B(H_1,\ldots,H_k)$ is either essential or inessential.

Thus $(H_1 \cup H_2') \cap H_3$, where H_2' is the complement $R^m - H_2$, is an essential set.

LEMMA 3. Let $H \subseteq R^m$ be an essential set, $Q(x)$ be a non-trivial polynomial such that $Q(x) \neq 0$ implies $x \in H$. If $b \in R^m$ satisfies $Q(b) \neq 0$ and $1 \leq k \leq m$, then there exists an $0 < \varepsilon$ such that for every $(c_{k+1},\ldots,c_m) = c$, $\|c - (b_{k+1},\ldots,b_m)\| < \varepsilon$ implies that the <u>section</u>

$$S = \{(a_1,\ldots,a_k) \mid (a_1,\ldots,a_k,c_{k+1},\ldots,c_m) \in H\},$$

is an essential set in R^k.

Here $||a||$ denotes the usual euclidean norm of the vector $\underline{a}$.

4. ALGORITHMS BY TESTS

We want to describe the most general algorithm for solving the equations (1) by means of tests. The notions presented here, as well as Theorem 4 concerning such algorithms, have wider applicability than the present context.

The first step of the algorithm is a test $a_{i_1} \cdot v_1 = \alpha_1$ where $1 \leq i_1 \leq n$ and $v_1 \in R^n$ is a fixed vector. The next coefficient vector a_{i_2} to be tested and the v_2 used are functions of α_1. Thus $i_2 = i_2(\alpha_1)$, $v_2 = (v_{2,1}(\alpha_1), \ldots, v_{2,n}(\alpha_1))$. Let $a_{i_1} \cdot v_1 = \alpha_1$, $a_{i_2} \cdot v_2 = \alpha_2, \ldots, a_{i_j} \cdot v_j = \alpha_j$, be the first j tests. The $j+1$-th test depends on $\alpha_1, \ldots, \alpha_j$. Thus $i_{j+1} = i_{j+1}(\alpha_1, \ldots, \alpha_j)$, $v_{j+1} = (v_{j+1,1}(\alpha_1, \ldots, \alpha_j), \ldots, v_{j+1,n}(\alpha_1, \ldots, \alpha_j))$. After k tests, the solution $x \in R^n$ of (1) is computed from $\alpha_1, \ldots, \alpha_k$. Thus $x = (x_1(\alpha_1, \ldots, \alpha_k), \ldots, x_n(\alpha_1, \ldots, \alpha_k))$.

We now make the limiting assumption, to be retained throughout the rest of the paper, that all the functions of $\alpha_1, \ldots, \alpha_k$ used in the algorithm are <u>rational functions</u>; i.e. the computations with these numbers involve just the rational operations. We could prove our theorems under more general assumptions concerning the functions used. But from the point of view of actual computer-algorithms, the restriction to rational functions covers all possible algorithms.

With the above restriction, we have to say how the indices $i_2, i_3, \ldots,$ are determined.

At the $j+1$-th step, rational functions $f_{j,1}(\alpha_1, \ldots, \alpha_j), \ldots, f_{j,m}(\alpha_1, \ldots, \alpha_j)$, are computed. Let $\mathrm{sgn}(f_{j,i}(\alpha)) = \varepsilon_j$, where $\mathrm{sgn}(x) = -1, 0, 1$, according as to whether $0 < x$, $x = 0$, or $x < 0$. As part of the algorithm we have a function $\varphi_j(\varepsilon_1, \ldots, \varepsilon_m) = i$ taking values in $\{1, \ldots, n-1\}$ so that $i_{j+1} = \varphi_j(\varepsilon_1, \ldots, \varepsilon_m)$. Also, the sequence $(\varepsilon_1, \ldots, \varepsilon_m)$ determines <u>which</u> rational functions $(v_{j+1,1}(\alpha), \ldots, v_{j+1,n}(\alpha))$ will be used to form the vector v_{j+1} used in the test $a_{i_{j+1}} \cdot v_{j+1} = \alpha_{j+1}$. Thus the vectors $v_1, v_2, \ldots,$ used in the tests are piecewise rational functions of $\alpha_1, \alpha_2, \ldots$. Note that this is, in fact, the case with the algorithm in section 2.

5. LINEARIZATION OF ALGORITHMS

We see that the most general form of an algorithm for solving (1) by means of tests has the form of a tree. The following theorem shows that each tree-like algorithm contains paths so that the computation along these linear paths gives the correct result in almost all cases. This "elimination of branching theorem" is a special case of a more general result which we shall not state here.

THEOREM 4. Let A be an algorithm of the form given in section 4 for solving (1) by means of tests. If A solves (1) essentially in $R^{n(n-1)}$ (see Definition 1), then there exists a path

(4) $a_{i_1} \cdot v_1 = \alpha_1, \ldots, a_{i_k} \cdot v_k = \alpha_k$

(5) $x = (x_1(\alpha_1, \ldots, \alpha_k), \ldots, x_n(\alpha_1, \ldots, \alpha_k))$,

of the tree, where $v_{j+1} = v_{j+1}(\alpha_1, \ldots, \alpha_j)$, i.e. each coordinate is a rational function of $\alpha_1, \ldots, \alpha_j$, such that, essentially in $R^{n(n-1)}$, (5) is a solution of (1). Thus this path is a linear algorithm for solving (1). Each i_{j+1}, and the choice of the rational functions $v_{j+1}(\alpha)$, do <u>not</u> depend on the previous $\alpha_1, \ldots, \alpha_j$.

PROOF. Let $f_1(\alpha_1, \alpha_2 \ldots)$, $f_2(\alpha_1, \alpha_2, \ldots), \ldots, f_p(\alpha_1, \alpha_2, \ldots)$, be a listing of all the rational functions $f_{j,\ell}(\alpha)$, used in determining the branchings in A. A moments reflection will show that along every fixed path of A, $\alpha_1, \alpha_2, \ldots$ are fixed rational functions in all the coordinates $a_{i,j}$ of $a = (a_1, \ldots, a_{n-1})$. Thus $f_j(\alpha) = F_j(a)$, $1 \leq j \leq p$, where each $F_j(a)$ is a rational function. We may assume that none of the $F_j(a)$ vanishes identically. Let $b \in R^{n(n-1)}$ be a point where none of the $F_j(b)$ vanishes. Let $0 < \varepsilon$ be such that $\|a-b\| < \varepsilon$ implies $\operatorname{sgn}(F_j(a)) = \operatorname{sgn}(F_j(b))$, $1 \leq j \leq p$. Then for all $a \in R^{n(n-1)}$ such that $\|a-b\| < \varepsilon$, the computation by the algorithm A will follow along the same path, say the path (4), (5). Now, in (5) the vector x also depends rationally on $a = (a_1, \ldots, a_{n-1})$, i.e. $x = x(a)$. Thus for all <u>a</u> such that $\|a-b\| < \varepsilon$, $a_i \cdot x(a) = 0$, $1 \leq i \leq n-1$. This implies that these equations hold identically in $R^{n(n-1)}$. Hence the linear algorithm (4), (5) will solve the equations (1) provided all the rational functions $v_{j,\ell}(\alpha)$ and $x_j(\alpha)$ are meaningful, i.e. none of the denominators vanishes. Thus the exceptional set is inessential.

6. REDUCTION OF THE PROBLEM

Henceforth let (4), (5) be an algorithm AL which, essentially in $R^{n(n-1)}$, solves (1). We want to prove $\frac{n(n+1)}{2} - 1 \leq k$.

LEMMA 5. Let $a, v_1, \ldots, v_m \in R^n$, denote $D = \{v_1, \ldots, v_m\}$, $\langle D \rangle = \langle v_1, \ldots, v_m \rangle$. If $u \notin \langle D \rangle$ then for every $0 < \varepsilon$ there exists an $a' \in R^n$ such that $a' \cdot v_i = a \cdot v_i$, $1 \leq i \leq m$, and $a' \cdot u \neq a \cdot u$.

PROOF. Decompose $u = v + d$, where $v \in \langle D \rangle$ and $d \perp D$. Then $d \neq 0$. Put $a' = a + \lambda d$, choosing $0 < \lambda < \varepsilon \, \|d\|^{-1}$

For the fixed algorithm AL given by (4), (5), and for $1 \leq i \leq n-1$, define

(6) $$D_i = \{v_j \mid i_j = i\},$$

the set D_i consists of all vectors v_j with which a_i was tested in (4).

THEOREM 6. If AL is, essentially in $R^{n(n-1)}$, an algorithm for solving (1) then, essentially in $R^{n(n-1)}$,

(7) $$N(a_1) \cap \langle D_1 \rangle \cap \ldots \cap N(a_{n-1}) \cap \langle D_{n-1} \rangle = \langle x \rangle$$

where x is the solution (5) of (1).

PROOF. We have $x \in N(a_i)$, $1 \leq i \leq n-1$. Let $Q(y_1, \ldots, y_{n(n-1)})$ be the polynomial such that $Q(a) \neq 0$ implies for $a = (a_1, \ldots, a_{n-1}) \in R^{n(n-1)}$ that AL does give a solution for (1). Let a $R^{(n-1)}$ satisfy $Q(a) \neq 0$. Assume that (7) does not hold, then for some D_i, say D_1, $x \notin \langle D_1 \rangle$. Take $0 < \varepsilon$ so small that $\|a - a'\| < \varepsilon$ implies $Q(a') \neq 0$. By Lemma 5, choose an $a'_1 \in R^n$ so that $\|a_1 - a'_1\| < \varepsilon$, $a'_1 \cdot v_j = a_1 \cdot v_j$ for $v_j \in D_1$, and $a'_1 \cdot x \neq a_1 \cdot x = 0$. Put $a' = (a'_1, a_2, \ldots, a_{n-1})$, then $\|a - a'\| = \|a_1 - a'_1\| < \varepsilon$. Hence AL solves (1) for the coefficient vectors $a'_1, a_2, \ldots, a_{n-1}$.

We come now to the main point of the proof. Since $a'_1 \cdot v_j = a_1 \cdot v_j$ for $v_j \in D_1$, it follows, by induction on the sequence (4), that the tests on $a'_1, a_2, \ldots, a_{n-1}$, produce the same values $\alpha_1, \ldots, \alpha_k$, as the tests on $a_1, a_2, \ldots, a_{n-1}$. Hence the computation (5) of the solution vector x from $\alpha_1, \ldots, \alpha_k$ will produce the same vector $x = (x_1, \ldots, x_n)$ in both cases. But $a'_1 \cdot x \neq 0$ contrary to the fact that AL produces a solution of (1) for $a'_1, a_2, \ldots, a_{n-1}$.

Thus if $Q(a) \neq 0$, then the solution x of (1) satisfies $x \in \langle D_i \rangle$, $1 \leq i \leq n-1$. Now (7) follows from $\bigcap_i N(a_i) = \langle x \rangle$.

COROLLARY 7. If AL given by (4), (5), solves (1) essentially in $R^{n(n-1)}$, then

(8) $$\dim(N(a_1) \cap \langle D_1 \rangle \cap \ldots \cap N(a_{n-1}) \cap \langle D_{n-1} \rangle) = 1,$$

holds essentially in $R^{n(n-1)}$.

To derive from (8) that $\frac{n(n+1)}{2} - 1 \leq k$, we consider a more general form of (8).

THEOREM 8. Let $S \subseteq R^n$ be a fixed subspace, $\dim(S) = p$. If the sequence of tests (4) on vectors $a_1, \ldots, a_m \in R^n$ (i.e. each i_j satisfies $1 \leq i_j \leq m$), where again each $v_{j+1} = v_{j+1}(\alpha_1, \ldots, \alpha_j)$ depends rationally on $\alpha_1, \ldots, \alpha_j$, satisfies for some $q \leq m$, essentially in R^{nm},

(9) $$\dim(S \cap N(a_1) \cap \langle D_1 \rangle \cap \ldots \cap N(a_q) \cap \langle D_q \rangle) = p-q,$$

then for some $i \leq q$, $p \leq c(D_i)$. Here $c(D)$ denotes the cardinality of the set D, so that a_i was tested in (4) at least p times.

PROOF. By induction on q. Let $q=1$ so that for some polynomial $Q(y)$ of mn variables, $Q(a) \neq 0$ implies $\dim(N(a_1) \cap \langle D_1 \rangle) = p-1$. By Lemma 3, find numerical vectors $c_2, \ldots, c_m \in R^n$ so that for $a_1, c_2, \ldots, c_m$, the previous equation holds essentially for $a_1 \in R^n$. In the sequence (4), every test $a_{i_j} \cdot v_j$ where $2 \leq i_j$ can now be dropped because $a_{i_j} = c_{i_j}$ is a fixed numerical vector so that $\alpha_j = c_{i_j} \cdot v_j$ can be directly computed from $\alpha_1, \ldots, \alpha_{j-1}$, which determine v_j, and from c_{i_j}. Assuming that (4) is already thus rewritten, it has the form

$$a_1 \cdot v_1 = \alpha_1, \ldots, \quad a_1 \cdot v_k = \alpha_k,$$

where v_1 is the first non-zero vector, and hence $D_1 = \{v_1, \ldots, v_k\}$. If $c(D_1) < p$ then we must have, $\dim(\langle D_1 \rangle) = p-1$, $S \cap N(a_1) \cap \langle D_1 \rangle = \langle D_1 \rangle$, hence $\langle D_1 \rangle \subseteq N(a_1)$, and $a_1 \cdot v_1 = 0$. But v_1 is a fixed vector so that $\{a_1 | a_1 \cdot v_1 = 0\}$ is an inessential set, contrary to $\dim(S \cap N(a_1) \cap \langle D_1 \rangle) = p-1$ holding essentially in R^n.

Induction step. Assume the statement to hold for $q-1$. Let (4) be a shortest sequence of tests so that (9) holds essentially in R^{nm}, but $c(D_i) < p$, $1 \leq i \leq q$. We must have $m=q$, for otherwise by specializing $a_{q+1}, \ldots, a_m$ in the manner of the previous paragraph, we would shorten the sequence (4). Hence

$1 \leq i_k \leq q$, say $i_k = q$ and $v_k \in D_k$.

Consider the sequence of tests

(10) $$a_{i_1} \cdot v_1 = \alpha_1, \ldots, a_{i_{k-1}} \cdot v_{k-1} = \alpha_{k-1}$$

obtained from (4) by dropping the last test $a_q \cdot v_k$. The D-sets corresponding to this sequence are $D_1, \ldots, D_{q-1}, D'_q = D_q - \{v_k\}$. Denote for $a = (a_1, \ldots, a_q) \in R^{nq}$, and for (4) and (10):

$$H(a) = S \cap N(a_1) \cap \langle D_1 \rangle \cap \ldots \cap N(a_q) \cap \langle D_q \rangle,$$

$$H'(a) = S \cap N(a_1) \cap \langle D_1 \rangle \cap \ldots \cap N(a_q) \cap \langle D'_q \rangle,$$

$$G(a) = S \cap N(a_1) \cap \langle D_1 \rangle \cap \ldots \cap N(a_{q-1}) \cap \langle D_{q-1} \rangle.$$

For essentially all $a = (a_1, \ldots, a_q) \in R^{nq}$, $\dim(H'(a)) = p-q$ or $\dim(H'(a)) = p-q-1$. Now the statement $\dim(H'(a)) = r$, where r is a fixed integer, is equivalent to a boolean combination of statements concerning vanishing of polynomials in $a \in R^{nq}$. Therefore, by Lemma 2, the set of $a \in R^{nq}$ for which $\dim(H'(a)) = r$ is either essential or inessential. Combining with the above, either essentially in R^{nq}, $\dim(H'(a)) = p-q$, or essentially $\dim(H'(a)) = p-q-1$. The first alternative implies that (10) is a sequence of tests for which (9) holds essentially, contrary to the minimality assumption on (4). Thus $\dim(H'(a)) = p-q-1$ holds essentially.

Similarly, $\dim(G(a)) = p-q+1$ or $\dim(G(a)) = p-q$, and one of these holds essentially. The first alternative implies, by the induction hypothesis, that $c(D_i) \geq p$ for some $1 \leq i \leq q-1$, thus finishing the proof. Otherwise $\dim(G(a)) = p-q$ holds essentially.

Hence, essentially in R^{nq}, $H(a) = G(a)$, and $H'(a) \subset H(a)$, $H'(a) \neq H(a)$. Let $a = (a_1, \ldots, a_q)$ be a point interior to the domain where all these statements hold. There is a vector u such that $u \in H(a)$, $u \notin H'(a)$. This implies $u \in G(a)$, $u \in N(a_q)$, and $u \notin \langle D'_q \rangle$. By Lemma 5, choose $a'_q \in R^n$ close enough to a_q so that $a' = (a_1, \ldots, a_{q-1}, a'_q)$ is also in the interior of the set where $H(a') = G(a')$ and $H'(a') \neq H(a')$, and also so that $a'_q \cdot v = a_q \cdot v$ for $v \in D'_q$, $a'_q \cdot u \neq a_q \cdot u = 0$.

Consider the sequence (4) for $a' = (a_1, \ldots, a_{q-1}, a'_q)$. From $a'_q \cdot v = a_q \cdot v$ for $v \in D'_q$, it follows, by induction on the sequence, that all the first $k-1$ tests (the sequence (10)) produce the same $\alpha_1, \ldots, \alpha_{k-1}$. Thus $D_1, \ldots, D_{q-1}$ are the same for a' as for $\underline{a}$, hence $G(a') = G(a)$. This implies, by $H(a') = G(a')$ and $H(a) = G(a)$, that $H(a') = H(a)$. Hence

$u \in H(a')$ so that $u \in N(a'_q)$. But this contradicts $a'_q \cdot u \neq 0$. This completes the proof.

7. THE MAIN THEOREM

THEOREM 9. Let A be an algorithm for solving (1) by means of tests as defined in section 4. Assume A is essentially valid in $R^{n(n-1)}$. Then, essentially in $R^{n(n-1)}$, the computation proceeds along paths of the form (4), (5), of the computation tree so that one of the a_i is tested at least n times, another a_i is tested at least $n-1$ times, and so on down to 2. Thus, essentially in $R^{n(n-1)}$, the total number of tests exceeds $n(n+1)/2-1$.

PROOF. From Theorem 4, and its proof, it follows that, essentially in $R^{n(n-1)}$, the solution by A proceeds along paths (4), (5), so that, again essentially in $R^{n(n-1)}$, each linear path is a solution algorithm AL for (1).

By Corollary 7, for such an AL, (8) holds essentially in $R^{n(n-1)}$. By Theorem 8, with $S = R^n$, and $\dim(S) = n$, this implies that one a_i, say a_{n-1}, is tested n times.

Since $\dim(N(a_1) \cap \ldots \cap N(a_{n-1})) = 1$ whenever $a_1, \ldots, a_{n-1}$ are linearly independent, (8) implies that $\dim (N(a_{n-1}) \cap N(a_1) \cap \langle D_1 \rangle \cap \ldots \cap N(a_{n-2}) \cap \langle D_{n-2} \rangle) = 1$, essentially in $R^{n(n-1)}$. Using Lemma 3, specialize a_{n-1} to a numerical vector $c \in R^n$ so that, essentially in $R^{n(n-2)}$, $\dim(N(c) \cap N(a_1) \cap \ldots \cap \langle D_{n-2} \rangle) = 1$. As in the proof of Theorem 8, this specialization $a_{n-1} = c$ turns (4) into a sequence of tests on $a_1, \ldots, a_{n-2}$. Now, $\dim(N(c)) = n-1$ so that by Theorem 8, some a_i, say a_{n-2}, is tested in (4) $n-2$ times. And so on down to a_1. This completes the proof.

SIMPLE PROOFS OF LOWER BOUNDS FOR POLYNOMIAL EVALUATION †

Edward M. Reingold and A. Ian Stocks

Department of Computer Science
University of Illinois at Urbana-Champaign
Urbana, Illinois 61801

INTRODUCTION

The idea of establishing lower bounds on the number of arithmetic operations required to evaluate polynomials is due originally to Ostrowski (1954A). He showed that at least n multiplications and n additions/subtractions are required to evaluate n^{th} degree polynomials for $n \leq 4$. Since then, this result has been proved true for all nonnegative values of n [Belaga (1961A), Pan (1966A)]. Motzkin (1955A) introduced the idea of preconditioning; if the same polynomial is to be evaluated at many points, it may be reasonable to allow some free preprocessing of the coefficients. It has been shown [Motzkin (1955A), Belaga (1961A)] that even if this preconditioning is not counted then at least $\lceil n/2 \rceil$ multiplications/divisions and n additions/subtractions are necessary to evaluate n^{th} degree polynomials. Eve (1964A) and others have shown that this lower bound is almost achievable: an nth degree polynomial can be evaluated in $\lfloor n/2 \rfloor + 2$ multiplications and n additions/subtractions, provided some irrational preconditioning is allowed without cost.

The previous proofs of the lower bounds for polynomial evaluation (with or without preconditioning) are not easy to follow. When preconditioning is permitted, either the proofs use difficult "degrees of freedom" arguments (see, e.g., §4.6.4 of Knuth (1969A)) or they use fairly deep results from topology [Winograd (1970A)]. When preconditioning is not allowed, the proofs are tedious since,

† This work was supported in part by the National Science Foundation (Grants GJ-31222 and GJ-812).

for the proof to work, a much stronger result must be stated and proved. The object of this paper is to present relatively simple, self-contained proofs of the above mentioned results. In particular, for preconditioning we do not use either degrees of freedom or results from topology, but rather algebraic independence, a simple and widely known concept. The other results are proved without substantially strengthening the statement of the result.

SCHEMES

In order to establish lower bounds for the evaluation of polynomials, it is necessary to specify a definite model of computation. The model we will use is the one usually chosen: a <u>scheme</u> is a sequence of steps

$$\lambda_i = \lambda_j \circ \lambda_k \qquad i > j,k \geqq 0 \tag{1}$$

in which "$\circ$" is one of the operations addition, subtraction, or multiplication.† Let the polynomial to be evaluated be

$$u_n x^n + u_{n-1} x^{n-1} + \ldots + u_1 x + u_0 \,.$$

The values of x, the coefficients, and arbitrary rational constants are introduced into the scheme by defining $\lambda_0 = x$, $\lambda_{-j-1} = u_j$, and $\lambda_{-n-1-i} = c_i$.

A scheme with preconditioning is a scheme as above, but instead we define $\lambda_{-j-1} = \alpha_j$ where the α_j are "parameters," $\alpha_j = f_j(u_n, \ldots, u_0)$ for arbitrary functions f_j. Eve's evaluation scheme uses irrational functions f_j; Rabin and Winograd (personal communication) have considered the case when the f_j are restricted to being rational functions.

In such schemes certain of the operations are, in some sense, more difficult than others. The step λ_i from (1) is called a <u>crucial step</u> (or <u>crucial operation</u>) in u_n provided that at least one of λ_j or λ_k is either u_n or is a crucial step in u_n and the other is not a constant.

†The proofs for preconditioning readily generalize to include division; the proof of Theorem 2 does not.

LOWER BOUNDS

The first two theorems show that at least n multiplications and n additions/subtractions are necessary to evaluate a general n^{th} degree polynomial. Thus Horner's rule is the optimum way to evaluate polynomials.

Theorem 1: Any scheme without preconditioning which evaluates a general polynomial of the form

$$U(x) = u_n x^n + u_{n-1} x^{n-1} + \dots + u_{n-p} x^{n-p} \qquad 0 \leq p \leq n$$

includes at least p addition/subtraction steps.

Proof: By induction on p, for p = 1 the result is obvious. Suppose that the theorem is true for all i, $i < p \leq n$. Let $\Lambda = \{\lambda_i\}$ be a scheme which correctly evaluates U(x) for any values of x, $u_n, u_{n-1}, \dots, u_{n-p}$. If there is no addition step in Λ which is crucial in u_{n-p} then the scheme correctly evaluates U(x) only if U(x) is a multiple of u_{n-p} or is independent of u_{n-p}. In either case, the scheme could not correctly evaluate all U(x) of the given form. Thus some addition step is crucial in u_{n-p} and setting $u_{n-p} = 0$ obviates the first such step. Thus with one fewer addition step the scheme evaluates

$$U(x) = u_n x^n + u_{n-1} x^{n-1} + \dots + u_{n-(p-1)} x^{n-(p-1)} .$$

which, by induction, requires p-1 addition steps. Thus Λ must have at least p addition steps. ∎

Theorem 2: There is no multivariate polynomial P for which the polynomial

$$U_P(x,u_n,\dots,u_0) = P(x,u_n,\dots,u_0)x^{n+1} + u_n x^n + u_{n-1} x^{n-1} + \dots + u_0$$

can be evaluated by a scheme with fewer than n crucial multiplications.

Proof: By induction on n, for n = 1 the result is obvious. Suppose that the theorem is true for all i < n. Let $\Lambda_P = \{\lambda_i\}$ be any scheme which correctly evaluates U_P for all values of x, u_n, ... u_0. If there is no crucial multiplication in Λ_P

involving u_n, then U_P must be linear in u_n or independent of it. In either case, the scheme could not correctly evaluate all U_P of the given form. Thus some multiplication step is crucial in u_n. Let

$$\lambda_\alpha = \lambda_\beta \cdot \lambda_\gamma, \quad \beta, \gamma < \alpha$$

be the first such multiplication. Since this is the first such multiplication, we can assume without loss of generality that

$$\lambda_\beta = c\, u_n + \lambda_\delta \ , \quad \delta < \beta$$

where $c \neq 0$ is a rational constant and λ_δ is of the form $f(x, u_{n-1}, \ldots, u_0)$, f a polynomial. The multiplication λ_α will be obviated if we force $\lambda_\beta = 0$ by considering inputs in which $u_n = -\frac{1}{c}\lambda_\delta$.

The steps prior to λ_α which involve u_n can only have the form

$$\lambda_i = c\, u_n + \lambda_j \ , \quad j < i \ . \tag{2}$$

Modify the scheme as follows: Change each step of the form (2) between λ_1 and λ_α to be $\lambda_i = \lambda_j$. Notice that none of the λ's which are thus changed can appear in a crucial multiplication before λ_α since λ_α was the first multiplication crucial in u_n. With these modifications, the step λ_β now calculates $\lambda_\beta = \lambda_\delta$. Delete this step and in its place put the step $\lambda_{-n-1} = -\frac{1}{c}\lambda_\delta$; this is <u>not</u> a crucial multiplication since c is a constant. After this step, insert steps to recompute each of the λ_i which were changed before; again, this involves no new crucial multiplications. The result of these modifications is a scheme with one less crucial multiplication which computes

$$U_{\hat{P}}(x, u_{n-1}, \ldots, u_0) = \hat{P}(x, u_{n-1}, \ldots, u_0)x^n + u_{n-1}x^{n-1} + \ldots + u_0 \ .$$

By induction, this requires at least n-1 crucial multiplications; thus Λ must have had n crucial multiplications. ■

<u>Corollary</u>: The evaluation of a general n^{th} degree polynomial requires n additions/subtractions and n multiplications.

<u>Proof</u>: Take p = n in Theorem 1 and P = 0 in Theorem 2. ■

We now consider the case in which preconditioning of the coefficients is allowed. For the sake of exposition we will initially consider schemes without division; the proofs generalize to include division, as we will subsequently show.

Lemma: Any scheme with preconditioning which evaluates a general polynomial

$$U(x) = u_n x^n + u_{n-1}x^{n-1} + \dots + u_0 .$$

uses at least n+1 algebraically independent parameters (the parameters are the results of the preconditioning which are given as inputs to the scheme).

Proof: Suppose there are only n parameters $\alpha_1, \dots, \alpha_n$. Then, by the definition of a scheme with preconditioning, the final result of the scheme must be of the form

$$f_n(\alpha_1,\dots,\alpha_n)x^n + f_{n-1}(\alpha_1,\dots,\alpha_n)x^{n-1} + \dots + f_0(\alpha_1,\dots,\alpha_n)$$

where each f_i is a polynomial in $\alpha_1,\dots,\alpha_n$ with rational coefficients. The n+1 functions $f_n,\dots,f_0$ cannot be algebraically independent (see §10.3 of Van der Waerden (1970A)) since they are polynomials with rational coefficients of only n parameters. On the other hand, the coefficients $u_n,\dots,u_0$ are clearly algebraically independent, a contradiction. A similar argument shows that at least n+1 of the parameters must be algebraically independent. ∎

Theorem 3: Any scheme with preconditioning which evaluates a general n^{th} degree polynomial includes at least $\lceil n/2 \rceil$ multiplication steps.

Proof: Notice that of the n+1 parameters, at most one can appear linearly in the result of the scheme; if more than one appears linearly, then a new parameter can be introduced so that it is the only one to appear linearly. Each of the other parameters is not linear and hence must appear in a multiplication step Each multiplication step can introduce at most two new parameters multiplicatively, and then only if the step has the form $\lambda_i = (\alpha_q + \lambda_j)(\alpha_r + \lambda_k)$. If additional parameters are introduced at that step without multiplications, then it must be additively, e.g. $(\alpha_p + \alpha_q + \lambda_i)(\alpha_r + \lambda_k)$; again, additional parameters can be introduced say, $\alpha = \alpha_p + \alpha_q$, in the preconditioning so that only two new parameters are introduced multiplicatively at each multiplication step. Thus, since the final scheme (resulting from the above modifications) must have at least n+1 parameters, only one of which can occur linearly, there must be at least $\lceil n/2 \rceil$ multiplication steps. ■

Theorem 4: Any scheme with preconditioning which correctly evaluates a general n^{th} degree polynomial includes at least n addition/subtraction steps.

Proof: For $n = 0$ the theorem is obvious. For $n > 0$ the proof will show that, after suitable modification, a scheme can introduce at most one parameter per additive step, except perhaps one parameter which can be introduced multiplicatively. Since, by the lemma, there are at least n+1 parameters, it will follow that there must be at least n addition/subtraction steps.

Let $\Lambda = \{\lambda_i\}$ be a scheme which correctly evaluates a general n^{th} degree polynomial. Let the set of parameters in the scheme be $\{\alpha_1, \ldots, \alpha_k\}$, $k \geq n+1$ assuming that steps of the form $\lambda = \alpha_i \circ \alpha_j$ have been eliminated by the introduction of new parameters. For each subset $Q \subseteq P$ construct Λ_Q, a variant of Λ, by assuming that $\alpha_i \in Q$ if and only if $\alpha_i = 0$ and then deleting all operations which have been made trivial.

We claim that at least one of the 2^k variant schemes, say $\Lambda_R = \{\lambda_i^R\}$, thus constructed must yield as its output a class of nth degree polynomials with n+1 algebraically independent coefficients. Suppose to the contrary that every such variant scheme has as its output polynomials with at most n algebraically independent coefficients. Thus for each of the 2^k polynomial forms (corresponding to the 2^k variant schemes) there is a non-trivial algebraic relation between the n+1 coefficients, in terms of the parameters. This gives us 2^k polynomials (in the coefficients of the arbitrary n^{th} degree polynomial to be evaluated) with the property that for every possible set of coefficients of the polynomial to be evaluated, one of those 2^k polynomials is zero. Hence the product of all 2^k polynomials is always zero and thus gives us a nontrivial algebraic relation between the coefficients of the arbitrary n^{th} degree polynomial to be evaluated, a contradiction.

To simplify the notation let $\Lambda_R = \{\lambda_i\}$, dropping the superscript R, and let the parameters of Λ_R be $\{\alpha_1, \ldots, \alpha_\ell\}$. Since

Λ_R contains no more additions than Λ, the theorem is proved if we show that Λ_R includes at least n addition steps.

Each of the addition steps in Λ_R can be rewritten as

$$A_i = P_1 a_1 + P_2 a_2$$

where P_1 and P_2 are products (possibly empty) of parameters while a_1 and a_2 are products (possibly empty) of previous A_j's. At least one of P_1 or a_1 is not empty as is one of P_2 or a_2 (otherwise the addition would be trivial).

Construct Λ_R', a scheme equivalent to Λ_R, with parameters P_i, which, for each addition

$$A_i = P_1 a_1 + P_2 a_2 \ ,$$

computes the products $P_1 a_1$, $P_2 a_2$ and does the addition. This scheme has the same number of additions as Λ_R (but possibly more multiplications) and it calculates the same set of polynomials. The only addition steps of Λ_R' which can introduce <u>two</u> of the parameters have the form

$$A = P_1 a_1 + P_2 a_2 \quad \text{or} \quad A = P_1 a_1 + P_2 \ .$$

Replace these steps by

$$\hat{A} = a_1 + \frac{P_2}{P_1} a_2 \qquad\qquad \hat{A} = a_1 + \frac{P_2}{P_1}$$

$$\text{or}$$

$$A = P_1 \hat{A} \qquad\qquad A = P_1 \hat{A}$$

respectively. Now, by "adjusting" multiplicatively each later occurrence of A we can delete the step $A = P_1 \hat{A}$. We now have a scheme in which the parameters are quotients of the parameters of Λ_R'; the quotients are well defined since none of the parameters can be zero, by construction. In this new scheme each addition introduces at most one parameter, and only one parameter may be introduced without an addition.

Since the scheme computes, by construction, a set of polynomials with n+1 algebraically independent coefficients, it must have at least n+1 parameters. Hence it must have at least n addition steps; but it has no more addition steps than our original Λ, so Λ must also have at least n addition steps. ∎

The proofs of two preceding theorems hold, with slight modification, even if division is permitted in the schemes.† The only change in the proofs of the theorems is that it is necessary to consider only "admissible" parameter sets -- those which do not cause division by zero. The lemma, however, needs a new proof since we cannot appeal to the algebraic dependence of the (rational) functions f_i of the parameters. The extension of the lemma to rational functions can be concluded from

Proposition: Any set of n+1 rational functions $f_0, f_1, \ldots, f_n$ of n parameters $\alpha_1, \alpha_2, \ldots, \alpha_n$ is algebraically dependent.

Proof: We will show that for some sufficiently large m there is a nonzero polynomial of degree m in n+1 variables which is identically zero on $f_0, \ldots, f_n$. Any polynomial of degree m in n+1 variables has $(m+1)^{n+1}$ different terms and hence that many independent coefficients β_j. Now, we substitute the f_i into a general m^{th} degree polynomial and put everything over a common denominator. Choose a k, independent of m, which is large enough so that the largest exponent of any α_j which appears in the numerator is less than or equal to mk. For example, k could be chosen as

$$k = \max_{1 \leq j \leq n} \sum_{i=0}^{n} (\text{power of } \alpha_j \text{ in the numerator of } f_i + \text{power of } \alpha_j \text{ in the denominator of } f_i)$$

Now this numerator, viewed a polynomial in the α_i, has no more than $(mk+1)^n$ coefficients, each of which is a linear combination of the coefficients β_j. Choosing m large enough so that

$$(m+1)^{n+1} > (mk+1)^n$$

† We must modify the definition of a crucial step in u_n to include division of a constant by u_n or by a λ_j which is crucial in u_n.

we can find a non-trivial set of coefficients β_j which cause the numerator to be zero, and hence there is a nontrivial polynomial in the f_i which is identically zero. ■

We can now conclude that the evaluation of a general n^{th} degree polynomial requires n additions/subtractions and $\lceil n/2 \rceil$ multiplications/divisions, even if preconditioning of the coefficients is allowed.

There is an immediate corollary which can be obtained:

Corollary: If (x_i) and (y_i) are general n vectors then computing their dot product requires at least n-1 additions/subtractions and $\lceil n/2 \rceil$ multiplications/divisions, even if separate preconditioning of the x_i and y_i is allowed.

Proof: Notice that the proofs of the previous two theorems hold even if we are given as input the set of values $\{1, x, x^2, \ldots, x^n\}$. Hence if there is a scheme which violates one of these lower bounds, apply it to the vectors $(a_1, a_2, \ldots, a_n)$ and $(x, x^2, x^3, \ldots, x^n)$ to compute $\sum_{i=1}^{n} a_i x^i$. Then, in one more addition we can compute $\sum_{i=0}^{n} a_i x^i$ and so violate one of the lower bounds already established. ■

The techniques used in the proofs of Theorems 3 and 4 can also be used to establish a generalization of the above corollary:

Theorem 5: If A is a general m by n matrix and x is a general n vector, then any scheme which computes their product Ax requires at least mn-m additions/subtractions and $\lceil mn/2 \rceil$ multiplications/divisions, even if preconditioning of A is allowed.

Proof: Any scheme which evaluates Ax for all values of x must involve at least mn parameters (the elements of A) since, by appropriate choice of x, we can determine all of the elements of A for a fixed A. Now, each parameter "used" must appear in a multiplication step since otherwise the scheme would produce the wrong answer in computing Ax where x = (0, 0, ..., 0). As in the argument in Theorem 3, there must then be at least $\lceil mn/2 \rceil$ multiplication/division steps. As to the number of addition/subtraction steps, since there are m outputs (the elements of Ax) at most m of the mn parameters can appear only multiplicatively. Then, by arguments similar to those in Theorem 4, it can be shown that there must be at least mn-m addition/subtraction steps. ■

The result in Theorem 5 that at least $\lceil mn/2 \rceil$ multiplications/divisions are required is due to Winograd (1970A) while the result on the number of additions/subtractions is, we believe, new.

ACKNOWLEDGMENT

We are indebted to Volker Strassen for pointing out to us that the generalization of the Lemma to include division requires, in addition to the Proposition, a proof that the functions $f_0, f_1, \ldots, f_n$ in the Lemma are indeed rational functions of the parameters. For the evaluation of polynomials, the proof is trivial; for the evaluation of rational functions it is not.

ON OBTAINING UPPER BOUNDS ON THE COMPLEXITY OF MATRIX MULTIPLICATION

Charles M. Fiduccia

Department of Computer Science

State University of New York at Stony Brook

ABSTRACT. For each matrix-matrix product AB there is an equivalent matrix-vector product Xy, with X of a special form. If the set of matrices of the form X is contained in a module generated by t matrices, each expressible as a column-row product cr, then t multiplications are sufficient to compute AB. The search for better algorithms for AB is reduced to the decomposition of X, thus circumventing the manipulation of products which appear in the final algorithm for AB.

INTRODUCTION

It is well known that two $n \times n$ matrices can be multiplied with n^3 multiplications. If $n = rs$, and two $m \times m$ matrices can be multiplied with $f(m)$ multiplications without using the commutative law, then by viewing an $n \times n$ matrix as an $r \times r$ matrix with $s \times s$ matrix entries, two $n \times n$ matrices can be multiplied with $f(r)f(s)$ multiplications [Winograd (1970A)]. For the 2×2 case, it has been shown the 7 multiplications are sufficient [Strassen (1969A)] and necessary [Hopcroft and Kerr (1971A), Winograd (1971A)] and thus that $7^{\log n} \approx n^{2.8}$ multiplications are sufficient to multiply two $n \times n$ matrices.

Strassen gives the following algorithm for the product

$$\begin{pmatrix} A & B \\ C & D \end{pmatrix} = \begin{pmatrix} a & b \\ c & d \end{pmatrix} \begin{pmatrix} \alpha & \beta \\ \gamma & \delta \end{pmatrix}$$

$m_1 \leftarrow (a-b)\alpha$	$m_7 \leftarrow (b+c)(\alpha-\delta)$
$m_2 \leftarrow b(\alpha+\gamma)$	
$m_3 \leftarrow (c-d)\delta$	$A \leftarrow m_1+m_2$
$m_4 \leftarrow c(\beta+\delta)$	$B \leftarrow m_5-m_7-m_1-m_4$
$m_5 \leftarrow (a+c)(\alpha+\beta)$	$C \leftarrow m_6+m_7-m_2+m_3$
$m_6 \leftarrow (b+d)(\gamma+\delta)$	$D \leftarrow m_4-m_3$

To verify this algorithm, one must expand the 7 product of sums into sums of products and then form the 4 indicated sums. The three-level structure (sums of products of sums) prevents verification by inspection, hindering the search for better algorithms for the n×n case. We will show that one need not consider products at all. For example, Strassen's method is equivalent to the decomposition (• stands for zero)

$$\begin{pmatrix} a & b & \cdot & \cdot \\ c & d & \cdot & \cdot \\ \cdot & \cdot & a & b \\ \cdot & \cdot & c & d \end{pmatrix} = \begin{pmatrix} a-b & \cdot & \cdot & \cdot \\ \cdot & \cdot & \cdot & \cdot \\ b-a & \cdot & \cdot & \cdot \\ \cdot & \cdot & \cdot & \cdot \end{pmatrix} + \begin{pmatrix} b & b & \cdot & \cdot \\ -b & -b & \cdot & \cdot \\ \cdot & \cdot & \cdot & \cdot \\ \cdot & \cdot & \cdot & \cdot \end{pmatrix} + \begin{pmatrix} \cdot & \cdot & \cdot & \cdot \\ \cdot & \cdot & \cdot & c-d \\ \cdot & \cdot & \cdot & \cdot \\ \cdot & \cdot & \cdot & d-c \end{pmatrix} + \begin{pmatrix} \cdot & \cdot & \cdot & \cdot \\ \cdot & \cdot & \cdot & \cdot \\ \cdot & \cdot & -c & -c \\ \cdot & \cdot & c & c \end{pmatrix}$$

$$+ \begin{pmatrix} \cdot & \cdot & \cdot & \cdot \\ \cdot & \cdot & \cdot & \cdot \\ a+c & \cdot & a+c & \cdot \\ \cdot & \cdot & \cdot & \cdot \end{pmatrix} + \begin{pmatrix} \cdot & \cdot & \cdot & \cdot \\ \cdot & b+d & \cdot & b+d \\ \cdot & \cdot & \cdot & \cdot \\ \cdot & \cdot & \cdot & \cdot \end{pmatrix} + \begin{pmatrix} \cdot & \cdot & \cdot & \cdot \\ b+c & \cdot & \cdot & -b-c \\ -b-c & \cdot & \cdot & b+c \\ \cdot & \cdot & \cdot & \cdot \end{pmatrix}$$

of the tensor product $I_2 \otimes \begin{pmatrix} a & b \\ c & d \end{pmatrix}$ into the sum of 7 matrices, each capable of being multiplied by a vector with a single multiplication. The above decomposition can be easily verified by inspection, thus simplifying the search for better matrix multiplication algorithms.

PRELIMINARIES

In this section, we introduce some notation and show that it is sufficient to consider matrix-vector multiplication.

For any nonempty set S, let $S^{m\times n}$ be the set for all m×n matrices over (with entries in) S. If A is a matrix over S, define the entry set E(A) of A to be the subset of S whose elements are the entries of A. We write $A \overset{E}{=} B$ iff E(A) = E(B). As there are infinitely many B such that $A \overset{E}{=} B$, the matrix A serves only to represent the set E(A).

Let $A = (a_{ij})$ be an m×n matrix and $B = (b_{ij})$ be an r×s matrix. The tensor product $A \otimes B = (a_{ij}B)$ is the mr×ns matrix obtained by replacing a_{ij} with the matrix $a_{ij}B$. The direct sum $A \oplus B$ is the (m+r)×(n+s) matrix $A \oplus B = \begin{pmatrix} A & O \\ O & B \end{pmatrix}$. If I_n is the n×n identity matrix, then $I_s \otimes A = A\oplus\ldots\oplus A$(s terms). For m=r and n=s, define

$A \cdot B = (a_{ij} b_{ij})$ to be the term-by-term product of A and B. If $b_1, \ldots, b_s$ are the s columns of B, define the rs-column vector $\kappa(B) = [b_1^T \ldots b_s^T]^T$, where $(\ )^T$ denotes matrix transpose. Finally, let e_i be the ith column of the identity matrix and e_i^T be the ith row.

LEMMA 1. Let B have s columns, $d = \kappa(B)$ and $C = I_s \otimes A$. Then $AB \overset{E}{\equiv} Cd$.

Proof.

$$AB = \begin{pmatrix} Ab_1 \ldots Ab_s \end{pmatrix} \overset{E}{\equiv} \begin{pmatrix} Ab_1 \\ \vdots \\ Ab_s \end{pmatrix} = \begin{pmatrix} A & & \\ & \ddots & \\ & & A \end{pmatrix} \begin{pmatrix} b_1 \\ \vdots \\ b_s \end{pmatrix} = Cd.$$

The above construction is representative of other similar constructions (See Examples 1 and 2). Thus if P and Q are permutation matrices of appropriate size, then $AB \overset{E}{\equiv} Cd \overset{E}{\equiv} (PCQ)(Q^{-1}d)$. Often, the form of A or B can be used to advantage. Thus if $d = Dc$, one may use $(CD)c$ to represent $E(AB)$. In view of Lemma 1, it is sufficient (for the purpose of computing $E(AB)$) to consider matrix-vector multiplication.

EXAMPLE 1.

$$\begin{pmatrix} a & b \\ c & d \end{pmatrix} \begin{pmatrix} w & y \\ x & z \end{pmatrix} \overset{E}{\equiv} \begin{pmatrix} a & b & \cdot & \cdot \\ c & d & \cdot & \cdot \\ \cdot & \cdot & a & b \\ \cdot & \cdot & c & d \end{pmatrix} \begin{pmatrix} w \\ x \\ y \\ z \end{pmatrix} \overset{E}{\equiv} \begin{pmatrix} a & \cdot & b & \cdot \\ \cdot & a & \cdot & b \\ c & \cdot & d & \cdot \\ \cdot & c & \cdot & d \end{pmatrix} \begin{pmatrix} w \\ y \\ x \\ z \end{pmatrix}$$

EXAMPLE 2.

$$\begin{pmatrix} a & b \\ c & d \end{pmatrix} \begin{pmatrix} w & x \\ x & z \end{pmatrix} \overset{E}{\equiv} \begin{pmatrix} a & b & \cdot & \cdot \\ c & d & \cdot & \cdot \\ \cdot & \cdot & a & b \\ \cdot & \cdot & c & d \end{pmatrix} \begin{pmatrix} w \\ x \\ x \\ z \end{pmatrix} = \begin{pmatrix} a & b & \cdot \\ c & d & \cdot \\ \cdot & a & b \\ \cdot & c & d \end{pmatrix} \begin{pmatrix} w \\ x \\ z \end{pmatrix} \overset{E}{\equiv} \begin{pmatrix} a & b & \cdot \\ \cdot & a & b \\ c & d & \cdot \\ \cdot & c & d \end{pmatrix} \begin{pmatrix} w \\ x \\ z \end{pmatrix}$$

BILINEAR CHAINS

In this section, we consider the computation of the matrix-vector product Xy by specialized straight-line algorithms, which we call bilinear chains, and show that a bilinear chain for Xy corresponds to a decomposition of X.

Let R be a ring* and K be a subring of the center of R, so that ar=ra for all (a,r) in K×R. There exists at least one such K, namely the subring generated by 1. For each fixed a in K, the unary operation $f_a(r) = ar$, mapping R into itself, is called a

*We assume that all rings have a unity.

K-dilation of ratio a. We do not view a K-dilation as a multiplication, reserving the latter term for the binary operation $(r,s) \to rs$ mapping $R \times R$ into R.

Let $X = (x_{ij})$ be a matrix variable which ranges over a nonempty subset S of $R^{m\times n}$ and $y = (y_1 \ldots y_n)^T$ be a vector variable which ranges over $R^{n\times 1} = R^n$. We seek an algorithm ϕ to compute $E(Xy)$ from $E(X) \cup E(y)$ for any pair (X,y) in $S \times R^n$. We confine our attention to algorithms which we call bilinear chains.

For convenience of notation, rename the entries of X so that $E(X) = \{x_1,\ldots,x_s\}$ and $E(y) = \{y_1,\ldots,y_n\}$, and let $x = [x_1,\ldots,x_s]^T$. Elements of $E(X)$, $E(y)$ and $E(Xy)$ are considered to be equal iff they are equal (in R) for all X in S, for all y in R^n and for all (X,y) in $S \times R^n$ respectively. Define $L_K(E(X))$ to be the set of all linear combinations $\sum_{i=1}^{s} a_i x_i$ with fixed coefficients in K.

DEFINITION 1. A K-chain ϕ for $E(Xy)$ is a finite sequence $\phi_1, \phi_2, \ldots$ such that for each z in $E(Xy)$ there is a k such that $\phi_k = z$, for all (X,y) in $S \times R^n$, where each ϕ_k is either a data step $\phi_k \leftarrow \delta$ with δ in $E(X) \cup E(y)$, or is obtained from previous steps $(\phi_i, \phi_j \quad i,j<k)$ by a K-dilation $\phi_k \leftarrow f_a(\phi_i)$, addition/subtraction $\phi_k \leftarrow \phi_i \pm \phi_j$, or multiplication $\phi_k \leftarrow \phi_i \phi_j$. The chain ϕ is called K-bilinear iff all multiplication steps are such that ϕ_i is in $L_K(E(X))$ and ϕ_j is in $L_K(E(y))$.

Since the usual method for matrix multiplication constitutes a bilinear chain, there exists a bilinear chain for $E(Xy)$.

LEMMA 2. Let ϕ be a K-chain for $E(Xy)$ with t multiplication steps $k_1,\ldots,k_t$. Then every step ϕ_i of ϕ is in $L_K(E(X) \cup E(y) \cup \{\phi_{k_1},\ldots,\phi_{k_t}\})$; that is, ϕ_i is of the form

$$\phi_i = \sum_{j=1}^{t} c_{ij}\phi_{k_j} + \sum_{j=1}^{n} d_{ij}y_j + \sum_{j=1}^{s} f_{ij}x_j \qquad \text{for all } (X,y) \text{ in } S \times R^n,$$

Proof. Data steps $x_1,\ldots,x_s,y_1,\ldots,y_n$ and multiplication steps $\phi_{k_1},\ldots,\phi_{k_t}$ are of the required form. All other steps are obtained from these by K-dilations, additions or subtractions and are therefore of the required form since K is stable under these operations.

THEOREM 1. There exists a K-bilinear chain ϕ for $E(Xy)$ with t multiplication steps iff there exist fixed matrices B, C, D over K and a $t\times t$ diagonal matrix U over $L_K(E(X))$ such that $X-D = CUB$ for all X in S. Moreover, if 0 is in S then $D = 0$.

This theorem equates the search for a better K-bilinear chain for E(Xy) to the search for a better decomposition of the matrix X.

Proof. Assume that ϕ is a K-bilinear chain for E(Xy) with t multiplication steps. Let the t multiplications be $u_1v_1,\ldots,u_tv_t$ and define

$$u = \begin{pmatrix} u_1 \\ \vdots \\ u_t \end{pmatrix} \qquad v = \begin{pmatrix} v_1 \\ \vdots \\ v_t \end{pmatrix} \qquad U = \begin{pmatrix} u_1 & & \\ & \ddots & \\ & & u_t \end{pmatrix}$$

Since ϕ is K-bilinear, there exist fixed matrices A in $K^{t\times s}$ and B in $K^{t\times n}$ such that u = Ax for all X in S and v = By for all y in R^n. Then Uv = UBy = $[u_1v_1\ldots u_tv_t]^T$ = u·v. By Lemma 2, there exist fixed matrices C in $K^{m\times t}$, D in $K^{m\times n}$ and F in $K^{m\times r}$ such that Xy = C(UBy) + Dy + Fx for all (X,y) in $S\times R^n$. The matrices A, B, C, D and F are easily obtained from ϕ. Let y range over the subset $\{0,e_1,\ldots,e_n\}$ of R^n, where e_i is the ith column of the n×n identity matrix I. y=0 gives Fx = 0, whence Xy = (CUB+D)y. $y=e_i$, $1\le i\le n$, gives XI = X = (CUB+D)I = CUB+D. Moreover, if 0 is in S, then D=0 since X=0 implies x=0 implies u=0 implies U=0.

Assume X-D = CUB, where B, C and D are fixed matrices over K and U is a t×t diagonal matrix over $L_K(E(X))$. Say

$$U = \begin{pmatrix} u_1 & & \\ & \ddots & \\ & & u_t \end{pmatrix}, \text{ and define } u = \begin{pmatrix} u_1 \\ \vdots \\ u_t \end{pmatrix}.$$ Since U is over $L_K(E(X))$, there exists a fixed matrix A in $K^{t\times r}$ such that u = Ax. Compute Xy = (CUB+D)y as C((Ax)·(By))+Dy.

Evidently, this constitutes a K-bilinear chain for E(Xy). The term-by-term product (Ax)·(By) = U(By) uses t multiplications, while all other computations can be carried out with K-dilations and addition/subtractions. For any particular decomposition of X, each of the six subproblems should be analyzed for optimality. If X is to be multiplied by several vectors, then u = Ax need only be computed once and may be viewed as a preconditioning of X.

EXAMPLE 3. Strassen's method for 2×2 case:

$$\begin{pmatrix} a & b & \cdot & \cdot \\ c & d & \cdot & \cdot \\ \cdot & \cdot & a & b \\ \cdot & \cdot & c & d \end{pmatrix} = \begin{pmatrix} 1 & 1 & \cdot & \cdot & \cdot & \cdot & \cdot \\ \cdot & -1 & 1 & \cdot & \cdot & 1 & 1 \\ -1 & \cdot & \cdot & -1 & 1 & \cdot & -1 \\ \cdot & \cdot & -1 & 1 & \cdot & \cdot & \cdot \end{pmatrix} \begin{pmatrix} a-b & & & & & & \\ & b & & & & & \\ & & c-d & & & & \\ & & & c & & & \\ & & & & a+c & & \\ & & & & & b+d & \\ & & & & & & b+c \end{pmatrix} \begin{pmatrix} 1 & \cdot & \cdot & \cdot \\ 1 & 1 & \cdot & \cdot \\ \cdot & \cdot & \cdot & 1 \\ \cdot & \cdot & 1 & 1 \\ 1 & \cdot & 1 & \cdot \\ \cdot & 1 & \cdot & 1 \\ 1 & \cdot & \cdot & -1 \end{pmatrix}$$

COROLLARY 1-1. There exists a K-bilinear chain for $E(Xy)$ with 1 multiplication step iff there exist a fixed matrices D in $K^{m\times n}$, c (column) in $K^{m\times 1}$, b (row) in $K^{1\times n}$ and an element u in $L_K(E(X))$ such that $X-D = ucb$.

Proof. Since K is in the center of R, cub = ucb. Compute $Xy = (cub+D)y$ as $c((ax)(by))+Dy$, where $u = ax$ for some fixed a in $K^{1\times s}$.

EXAMPLE 4.

$$\begin{pmatrix} a & a \\ a & a \end{pmatrix} = a\begin{pmatrix} 1 & 1 \\ 1 & 1 \end{pmatrix} = a\begin{pmatrix} 1 \\ 1 \end{pmatrix}\begin{pmatrix} 1 & 1 \end{pmatrix} = \begin{pmatrix} 1 \\ 1 \end{pmatrix} a \begin{pmatrix} 1 & 1 \end{pmatrix} = a(e_1+e_2)(e_1+e_2)^T$$

$$\begin{pmatrix} -a & a \\ -a & a \end{pmatrix} = a\begin{pmatrix} -1 & 1 \\ -1 & 1 \end{pmatrix} = a\begin{pmatrix} -1 \\ -1 \end{pmatrix}\begin{pmatrix} 1 & 1 \end{pmatrix} = \begin{pmatrix} -1 \\ -1 \end{pmatrix} a \begin{pmatrix} 1 & 1 \end{pmatrix} = a(-e_1-e_2)(e_1+e_2)^T$$

$$\begin{pmatrix} \cos^2\theta & \sin^2\theta \\ \sin^2\theta & \cos^2\theta \end{pmatrix} = \begin{pmatrix} 1-\sin^2\theta & \sin^2\theta \\ \sin^2\theta & 1-\sin^2\theta \end{pmatrix} = \begin{pmatrix} -1 \\ 1 \end{pmatrix} \sin^2\theta \begin{pmatrix} 1 & -1 \end{pmatrix} + \begin{pmatrix} 1 & 0 \\ 0 & 1 \end{pmatrix}$$

As can be seen from example 3, the CUB+D decomposition of Theorem 1 explicitly shows the number of multiplication steps the resulting chain will have. However, it is difficult to directly decompose X into CUB+D and just as difficult to verify the decomposition by inspection. The following theorem suggests a simpler-but equivalent-decomposition.

THEOREM 2. There exists a K-bilinear chain for $E(Xy)$ with t multiplication steps iff there exist a fixed matrix D in $K^{m\times n}$ and t $m\times n$ matrices $X_1,\ldots,X_t$ each of the form $X_i = u_i c_i b_i$, where $c_1,\ldots,c_t$ are fixed column vectors in $K^{m\times 1}$, $b_1,\ldots,b_t$ are fixed row vectors in $K^{1\times n}$ and $u_1,\ldots,u_t$ are in $L_K(E(X))$, such that

$$X-D = \sum_{i=1}^{t} X_i \quad \text{for all X in S.}$$

Proof. In Theorem 1, let $c_1,\ldots,c_t$ be the columns of C; $b_1,\ldots,b_t$ be the rows of B; $u_1,\ldots,u_t$ be the diagonal entries of U and let $X_i = u_i c_i b_i$. Then

$$X-D = CUB = \sum_{i=1}^{t} c_i u_i b_i = \sum_{i=1}^{t} u_i c_i b_i = \sum_{i=1}^{t} X_i.$$

In the computation of $E(Xy)$, b_i serves to form the linear combination $b_i y$ in $L_K(E(y))$, u_i acts as a left multiplier $u_i(b_i y)$, while c_i serves to dilate (weigh) and distribute the product $u_i(b_i y)$ to each element of $E(Xy)$. The utility of theorem 2 stems from the fact that X may be decomposed a step at a time by subtracting from it each of the X_i. At each step, the decomposition is easily verified since only matrix addition is involved.

EXAMPLE 5.

$$\begin{pmatrix} a & \cdot \\ a\pm b & b \end{pmatrix} = \begin{pmatrix} a & \cdot \\ a & \cdot \end{pmatrix} + \begin{pmatrix} \cdot & \cdot \\ \pm b & b \end{pmatrix} = a(e_1+e_2)e_1^T + be_2(e_2\pm e_1)^T$$

EXAMPLE 6. A $\sum_{i=1}^{7} X_i$ decomposition of $X = I_2 \otimes \begin{pmatrix} a & b \\ c & d \end{pmatrix}$:

$$\begin{pmatrix} a & b & \cdot & \cdot \\ c & d & \cdot & \cdot \\ \cdot & \cdot & a & b \\ \cdot & \cdot & c & d \end{pmatrix} = \begin{pmatrix} a & a & \cdot & \cdot \\ a & a & \cdot & \cdot \\ \cdot & \cdot & \cdot & \cdot \\ \cdot & \cdot & \cdot & \cdot \end{pmatrix} + \begin{pmatrix} \cdot & \cdot & \cdot & \cdot \\ \cdot & \cdot & \cdot & \cdot \\ \cdot & \cdot & d & d \\ \cdot & \cdot & d & d \end{pmatrix} + \begin{pmatrix} \cdot & b-a & \cdot & \cdot \\ c-a & d-a & \cdot & \cdot \\ \cdot & \cdot & a-d & b-d \\ \cdot & \cdot & c-d & \cdot \end{pmatrix} = X_1+X_2+L \quad \text{(say)}$$

Note that by removing $\begin{pmatrix} a & a \\ a & a \end{pmatrix}$ from the first block of X and $\begin{pmatrix} d & d \\ d & d \end{pmatrix}$ from the second block of X, the result L necessarily has its (2,2) entry equal to the negative of its (3,3) entry. Further decompose L as follows:

$$\begin{pmatrix} \cdot & b-a & \cdot & \cdot \\ c-a & d-a & \cdot & \cdot \\ \cdot & \cdot & a-d & b-d \\ \cdot & \cdot & c-d & \cdot \end{pmatrix} = \begin{pmatrix} \cdot & \cdot & \cdot & \cdot \\ \cdot & d-a & a-d & \cdot \\ \cdot & d-a & a-d & \cdot \\ \cdot & \cdot & \cdot & \cdot \end{pmatrix} + \begin{pmatrix} \cdot & b-a & \cdot & \cdot \\ c-a & \cdot & d-a & \cdot \\ \cdot & a-d & \cdot & b-d \\ \cdot & \cdot & c-d & \cdot \end{pmatrix} = X_3+M \quad \text{(say)}$$

Note that M is the sum of 2 matrices, each having the form of the matrix of example 5, since a-d = (b-d)-(b-a) and d-a = (c-a)-(c-d). Thus $X = X_1+X_2+X_3+M$, where $M = X_4+X_5+X_6+X_7$ is the sum of 4 matrices each of the form $X_i = u_i c_i b_i$.

In order to obtain some insight into the number of multiplication steps which are necessary and sufficient in a K-bilinear chain for E(Xy), we take a new viewpoint of the results of theorem 2. In the decomposition $X-D = \sum_{i=1}^{t} X_i$, each X_i is of the form $X_i = u_i\xi_i$, where each $\xi_i = c_i b_i$ is fixed m×n matrix of rank 1 over the commutative subring K and each u_i is in $L_K(E(X))$. Thus $X-D = u_1\xi_1+\dots+u_t\xi_t$ for all X in S. It follows that the set S_D obtained by subtracting D from each element of S must be a subset of the R-module (vector space if R is a field) generated by the m×n matrices $\xi_1,\dots,\xi_t$. Conversely, if X ranges over any such subset, then there exists a K-bilinear chain for E(Xy) with t multiplication steps.

THEOREM 3. There exists a K-bilinear chain for E(Xy) with t multiplication steps iff there exist fixed vectors d in K^{mn}; $c_1,\dots,c_t$ in K^m; $b_1,\dots,b_t$ in K^n and elements $u_1,\dots,u_t$ in $L_K(E(X))$ such that $\kappa(X)-d = u_1(c_1\otimes d_1)+\dots+u_t(c_t\otimes d_t)$.

Proof. Take $d_i = b_i^T$ in theorem 2. Then apply κ to $X-D = \sum_{i=1}^{t} u_i c_i b_i$ to obtain $\kappa(X)-d = \sum_{i=1}^{t} u_i(c_i\otimes d_i)$, where $d = \kappa(D)$ and $\kappa(c_i b_i) = c_i\otimes b_i^T = c_i\otimes d_i$.

COROLLARY 3-1. If there exists a K-bilinear chain for $E(Xy)$ with t multiplication steps, then there exist a fixed vector d in $K^{mn\times 1}$, a fixed matrix M in $K^{mn\times t}$ and a t-vector u over $L_K(E(X))$ such that $\kappa(X)-d = Mu$.

Proof. Let M be the matrix whose i^{th} column is $c_i \otimes d_i$ and $u = [u_1,\ldots,u_t]^T$ be defined as theorem 1. Then

$$\sum_{i=1}^{t} u_i(c_i \otimes d_i) = \sum_{i=1}^{t} (c_i \otimes d_i) u_i = Mu.$$

Note that since $u = Ax$, then $\kappa(X)-d = MAx$. If $d=0$, it follows that M contains a submatrix which is a left inverse of A, since x is a subvector of $\kappa(X)$.

LOWER BOUNDS

In this section, we obtain several lower bounds on the number of multiplication steps necessary in a K-bilinear chain for $E(Xy)$. A duality result is also presented. We assume that R and K are nontrivial.

THEOREM 4. There is a K-bilinear chain for $E(Xy)$ with t multiplication steps iff there is a K-bilinear chain for $E(X^Tz)$ (where z ranges over R^m) with t multiplication steps.

Proof. By theorem 1, $X = CUB+D$ and thus $X^T = (CUB+D)^T = B^TUC^T+D^T$. Any K-bilinear chain for $E(Xy)$ can be transformed to a K-bilinear chain for $E(X^Tz)$. The transformation preserves multiplications and dilations but not additions/subtractions.

The following lower bound results follow from linear independence arguments and make use of the following

LEMMA 3. If A is a fixed $m\times n$ matrix over a commutative ring and $m>n$ $(n>m)$, then the rows (columns) of A are linearly dependent.

COROLLARY 1-2. If X has an $r\times c$ submatrix X' such that there exists no nontrivial fixed α in $K^{1\times r}$ for which $E(\alpha X')$ is a subset of K, then any K-bilinear chain for $E(Xy)$ has at least r multiplication steps.

Proof. Assume ϕ is a K-bilinear chain for $E(Xy)$ with $t<r$ multiplication steps. Then by theorem 1, $X = CUB+D$ and $X' = C'UB'+D'$, where C' is $r\times t$. By lemma 3, there exists a fixed nontrivial α in $K^{1\times r}$ such that $\alpha C' = 0$. Thus $\alpha X' = \alpha C'UB'+\alpha D' = \alpha D'$. But $\alpha D'$ is in $K^{1\times c}$, whence $E(\alpha X')$ is a subset of K--a contradiction.

Proofs for the following results are similar to the proof of corollary 1-2.

COROLLARY 1-3. (Winograd(1970A)) If X has an $r\times c$ submatrix X' such that there exists no nontrivial fixed β in $K^{c\times 1}$ for which $E(X'\beta)$ is a subset of K, then any K-bilinear chain for E(Xy) has at least c multiplication steps.

COROLLARY 1-4. (Fiduccia (1971A)) If X has an $r\times c$ submatrix X' such that there exist no nontrivial fixed α in $K^{1\times r}$ and β in $K^{c\times 1}$ for which $E(\alpha X'\beta)$ is a subset (singleton) of K, then any K-bilinear chain for E(xy) has at least r+c-1 multiplication steps.

COROLLARY 3-2. If there exists an e-vector x' over E(X) such that there exists no nontrivial fixed γ in $K^{1\times e}$ for which $E(\alpha x')$ is a subset (singleton) of K, then any K-bilinear chain for E(Xy) has at least e multiplication steps.

Proof. By corollary 3-1, $x' = M'u+d'$ for some fixed M' in $K^{e\times t}$ and d' in $K^{e\times 1}$. If t<e then, by lemma 3, $\gamma M' = 0$ for some nontrivial γ in $K^{1\times e}$. Then $\gamma x' = \gamma d'$ is in K--a contradiction.

UPPER BOUNDS

In this section we obtain upper bounds on several matrix multiplication problems by using the decompositions of theorems 1 and 2. Unless otherwise indicated, R is an arbitrary ring with unity 1 and K is the subring generated by 1. Thus all K-dilations can be replaced with additions/subtractions.

PROPOSITION 1. Two $n\times n$ matrices can be multiplied with $n^2-n(n-1)/2$ multiplications.

Sketch of proof. Replace the original problem AB by Xy, where $X = I_n \otimes A$. Decompose X as in example 6 by removing a block from X for each diagonal element of A. Any two blocks of the resulting matrix will have the form of L in example 6. That is the (j,j) entry of block i will be the negative of the (i,i) entry of block j, $1\le i<j\le n$. Proceed as in example 6 for each of these n(n-1)2 pairings, $1\le i<j\le n$. Each pairing removes one multiplication from the n^2 previously required.

PROPOSITION 2. Let X be $n\times n$ of the form $x_{ij} = \pm x_{ji}$. If ℓ is the number of entries x_{ij}, i<j, which are not identically zero, then $n+\ell$ multiplications are sufficient to compute Xy.

Proof by example. Consider the 3 3 case:

$$\begin{pmatrix} a & b & c \\ b & d & e \\ -c & e & f \end{pmatrix} = \begin{pmatrix} b & b & \cdot \\ b & b & \cdot \\ \cdot & \cdot & \cdot \end{pmatrix} + \begin{pmatrix} -c & \cdot & c \\ \cdot & \cdot & \cdot \\ -c & \cdot & c \end{pmatrix} + \begin{pmatrix} \cdot & \cdot & \cdot \\ \cdot & e & e \\ \cdot & e & e \end{pmatrix} + \begin{pmatrix} a-b+c & \cdot & \cdot \\ \cdot & d-b-e & \cdot \\ \cdot & \cdot & f-c-e \end{pmatrix}$$

For each $i<j$, if $x_{ij}\neq 0$ remove a block of the form shown in example 4. Each of these ℓ blocks uses one multiplication and introduced an entry along the main diagonal. Thus $n+\ell$ multiplications are sufficient.

COROLLARY 1. Two complex numbers can be multiplied with 3 real multiplications.

Proof. $(a+ib)(c+id) = \alpha+i\beta$, where $\begin{pmatrix} \alpha \\ \beta \end{pmatrix} = \begin{pmatrix} a & -b \\ b & a \end{pmatrix}\begin{pmatrix} c \\ d \end{pmatrix}$

Decompose the resulting matrix as follows:

$$\begin{pmatrix} a & -b \\ b & a \end{pmatrix} = \begin{pmatrix} a+b & \cdot \\ \cdot & a-b \end{pmatrix} + \begin{pmatrix} -b & -b \\ b & b \end{pmatrix}$$

COROLLARY 2. Two quaternions can be multiplied with 10 real multiplications.

Proof. The resulting matrix is given below and satisfies proposition 2 with $n=4$ and $\ell=6$.

$$\begin{pmatrix} a & -b & -c & -d \\ b & a & -d & c \\ c & d & a & -b \\ d & -c & b & a \end{pmatrix}$$

COROLLARY 3. An $m\times m$ matrix can be multiplied by an $m\times 2$ matrix (two vectors) with $(3m^2+5m)/2$ multiplications.

Proof. Replace the original product $A[b_1 b_2]$ by $\begin{pmatrix} \cdot & A \\ A & \cdot \end{pmatrix}\begin{pmatrix} b_1 \\ b_2 \end{pmatrix}$.

Then $\begin{pmatrix} \cdot & A \\ A & \cdot \end{pmatrix} = \begin{pmatrix} \cdot & A^T \\ A & \cdot \end{pmatrix} + \begin{pmatrix} \cdot & A-A^T \\ \cdot & \cdot \end{pmatrix}$.

The first matrix satisfies proposition 2 with $n=2m$ and $\ell=m^2$, while $A-A^T$ satisfies proposition 2 with $n=m$ and $\ell=m(m-1)/2$ for a total of $m^2+2m+m+m(m-1)/2 = (3m^2+5m)/2$ multiplications.

CONCLUSION

The search for better matrix multiplication schemes has been reduced to a matrix decomposition problem. In this new setting, formula manipulation is held to a minimum, while the intrinsic symmetries of the problem are accentuated. Theorem 3 shows that tensor products play a fundamental role in the analysis of the complexity of the matrix multiplication problem.

EFFICIENT ITERATIONS FOR ALGEBRAIC NUMBERS

Michael S.Paterson

Department of Computer Science
University of Warwick,Coventry,England

ABSTRACT

This paper is concerned with iterative procedures for finding real roots of rational polynomials, and in particular with notions of efficiency for such procedures. For a measure of efficiency, similar to Ostrowski's index but based on the number of elementary arithmetic operations used, the main theorem provides an upper bound. This bound is attained for quadratic polynomials by Newton's method. Some open problems and conjectures of a theoretical nature are presented.

1. INTRODUCTION

Many and varied are the iterative procedures which are commonly employed for finding zeroes of functions. It is natural to want to compare the efficacy and efficiency of different methods and to ask questions concerning optimality. Most previous research in this area [Traub (1964A), Traub (1971A), Ostrowski (1966A), Cohen and Varaiya (1970A), Feldstein and Firestone (1969A), Rissanen (1970A), Winograd and Wolfe (1971A)] has considered a very general problem and measured the amount of computation required to achieve a certain rate of convergence solely in terms of the number of times that the function or any of its derivatives are evaluated. Thus the results are most relevant for complicated functions where these evaluation times are large compared with the subsequent combination of the values obtained.

Frequently however the function concerned is a polynomial, in which case few of the previous results are directly useful. The computation of an iteration formula may perhaps never evaluate the polynomial or any derivatives explicitly, or, even if it does, this may be only a minor part of the total computation. In this paper the calculation of an iteration formula is regarded as an undemarcated sequence of rational operations. We set out to pose some of the questions which might be asked about the best kind of iteration algorithm to use for particular kinds of polynomial, and also more general questions such as whether it really does require more effort to approximate a zero for a polynomial of high degree than for a quadratic polynomial. As usual, few of these questions are answered conclusively, but we hope that some of these open problems may be taken up by more able people.

2. PRELIMINARIES

An informal style of presentation is to be adopted throughout the remainder of the paper because the details of the definitions and theorems are not really too important, but there are one or two ideas I should like to get across to you as easily as possible. I shall be concerned exclusively with iterative methods for approximating real algebraic numbers, that is,"finding" real simple zeroes of polynomials with rational coefficients.

The first definition we need is one for the rate of convergence or <u>order</u> of an iterative method. If a sequence $\{x_k\}$ converges to a limit α then the order of the sequence is the supremum of numbers p such that:

$$\overline{\lim_{k}} \frac{|x_{k+1} - \alpha|}{|x_k - \alpha|^p} < \infty$$

It happens that, for the class of iterative procedures I shall consider and for simple zeroes, the order of the sequence generated is independent of the polynomial, and that for some positive *integer* p

$$\lim_{k} \frac{x_{k+1} - \alpha}{(x_k - \alpha)^p} = C, \quad 0<|C|<\infty$$

The <u>order of</u> such <u>a procedure</u> is this integer p. Fortunately I do not have to worry about the difficulties of defining "order" for more general algorithms as do Cohen and Varaiya (1970A) and Rissanen (1970A), for example.

Suppose you have an iteration procedure with order p, then you get a new "improved" procedure of order p^2 or better by just doing two of the old iterations at a time. Of course this is absolutely no practical improvement as long as it takes twice as much computation to do two steps as it does to do one. Therefore to measure the efficiency of an iterative method we need to balance the order achieved against the "cost" per iteration. As I mentioned in the introduction other people have kept control of the cost by counting the number of times per iteration that the function or its derivatives are evaluated. Aside from the difficulty that this number may actually be zero, this measure takes no account of the fact that in many instances the cost of evaluating, for example, a function and its derivative simultaneously may be much less than the sum of the costs of evaluating each independently.

For this paper, I have avoided these difficulties by making the cost depend on the number of individual arithmetic operations required by the iteration algorithm. Therefore, unlike the order, it is dependent on the function whose zero is to be determined, and we can ask which method is best for a particular function. All of the usual iterative algorithms I can think of for solving polynomial equations with rational coefficients use only the rational operations, $\{+, -, \times, \div\}$, and rational constants. Of course if arbitrary constants were permitted in the present case there would be no need to iterate!

3. DEFINITIONS OF "COST"

A most reasonable definition of "cost", which must claim at least equal practicality with my eventual definition, is simply the total number of rational operations employed. I found this unsatisfactory for two reasons however. Firstly my main theorem comes out very much better with my definition. The second reason occupies the rest of this section.

The emphasis of nearly all theoretical work in this area is on the limiting behaviour of iterative processes, and certainly if the limit is to be arbitrarily closely approached the precision of the arithmetic operations must be indefinitely increased. For any kind of conventional computer implementation the time for arithmetic operations with n significant figures is at least linear in n for large enough n. With a fixed iterative procedure of order p (> 1) the number of significant figures in the result, and hence required in the arithmetic operations, increases by a factor of p at each iteration. Consequently the proportion of the total computation time spent on the final iteration alone is approximately $(1 - 1/p)$. This consideration has a marked effect on the criteria for efficient algorithms when very high precision is required. For the present I shall merely raise the following questions:

(i) to what extent is this "limiting behaviour" relevant to practical computations?

(ii) what are the most suitable kinds of algorithm under these conditions?

(iii) in particular, are there better ways to find successive digits of the limit than by arithmetic operations?

Maybe these matters are too far from everyday calculation to evoke useful results, but certainly there is frequent demand for a degree of approximation that requires multiple-precision arithmetic in at least the later stages of the computation. In multiple-precision arithmetic the cost of multiplications and divisions is increased relative to the cost of additions and subtractions. In most implementations the latter increases only linearly with the number of digits while the former grows with the square of the number of digits. Even the best algorithms known for n-digit multiplication and division require time proportional to n.log n.log log n [Schönhage and Strassen (1972A)]. Multiplication or division of an n-digit quantity by a single-precision number should however only need a time linear in n. This differentiation between linear and superlinear operations motivates my definition of "cost".

An algorithm Φ for an iteration function is, for me, a finite set of rational parameters, $\{\lambda_1, \ldots, \lambda_R\}$ and a finite sequence of rational functions $t_o = x, t_1, \ldots, t_T = \Phi(x)$ where for each $i \geqslant 1$, t_i has one of the following forms:

$$t_i = \begin{cases} \left.\begin{array}{l} t_j \times t_k \\ t_j \div t_k \\ \lambda_r \div t_k \end{array}\right\} & M \\ \left.\begin{array}{l} \lambda_r \times t_k \\ t_j \pm t_k \\ \lambda_r \pm t_k \end{array}\right\} & A \end{cases}$$

where $o \leqslant j,k < i$ and $1 \leqslant r \leqslant R$.

In the computation corresponding to such a sequence the first three forms will be regarded as needing superlinear arithmetic and are called M-operations, the remainder are A-operations. The cost of Φ, $M(\Phi)$, is the number of M-operations in the computation of Φ. I will also use Φ to denote the rational function computed by the algorithm, and though of course the same function may be computed in various ways I can suggest the intended algorithm by the expression used, for example:

$$\tfrac{1}{2}x - 3/x \qquad \text{or} \qquad (x^2 - 6)/2x$$

If I seem to have left out iterations with "memory" [see Traub (1964A)], just wait until the end of the next section.

4. "EFFICIENCY"

For an iteration algorithm Φ with order $p(\Phi)$ and cost $M(\Phi)$, how should the "efficiency" $\gamma(p,M)$ of Φ be defined? It should decrease as M increases and increase with p. Furthermore if $\Phi^{(n)}$ is the algorithm where we do n steps of Φ at a time, we should want:

$$\begin{aligned}\gamma(p,M) &= \gamma(p(\Phi^{(n)}), M(\Phi^{(n)})) \\ &= \gamma(p^n, nM)\end{aligned}$$

From this equation one can see that γ should depend only on $(\log p)/M$, and so we may as well take this as our definition.

Efficiency of $\Phi = \gamma(\Phi) = (\log_2 p(\Phi))/M(\Phi)$. Taking logarithms to base 2 is arbitrary but convenient at times. All my logarithms will be to base 2 unless specified otherwise. This definition of γ is not original, it is the logarithm of Ostrowski's efficiency index [Ostrowski (1966A)]. Note that I have not assumed that evaluating the *function* $\Phi^{(n)}$ takes n times the cost of evaluating Φ and there may indeed be cheaper ways of doing it. For examples, consider $\Phi(x) = 1/x$ or the first algorithm below. The point of the definition of efficiency is that if you want to gain by doing several iterations in one you have to think of some real labour-saving device.

A very well-known iterative procedure is the secant method where:

$$x_{i+1} = S(x_i, x_{i-1}) \equiv \frac{x_{i-1}\, f(x_i) - x_i f(x_{i-1})}{f(x_i) - f(x_{i-1})}$$

This is an example of an algorithm with "memory", which at first sight seems beyond my scope. The definitions of p, M and γ are completely straightforward for this case but from any algorithm with "memory" of efficiency e, we can easily derive algorithms without memory with efficiencies abritrarily close to e. For the secant method for instance define $\Phi_r(x)$ as follows:

$$x_o = x + f(x)$$

$$x_1 = x$$

$$x_{i+1} = S(x_i, x_{i-1}) \quad \text{for } i = 1, \ldots, r-1$$

$$\Phi_r(x) = x_r$$

Clearly as $r\to\infty$, $\gamma(\Phi_r) \to \gamma(S)$. To keep the discussion simple therefore, I only make definitions and prove theorems for memory-less procedures. I shall finish this section with three examples of iterative schemes for the equation:

$$f(x) \equiv x^2 + bx + c = 0$$

(I) Secant method doing two steps at a time with a "trick". The saving in cost is by using the two expressions

$$S(u,v) = v - (v^2 + bv + c)/(u + v + b) \equiv S_1(u,v)$$

$$= u - (u^2 + bu + c)/(u + v + b) \equiv S_2(u,v)$$

alternately.

Given the pair (u,v), compute $u' = S_1(u,v)$ using two M-operations. Using the sub-expression $(v^2 + bv + c)$ already computed, the computation of $v' = S_2(v,u')$ needs only one more M-operation.

Then $\Phi(u,v) = (u',v')$.

$p(\Phi) = \frac{1}{2}(3+\sqrt{5}) \doteq 2.618$; $M(\Phi) = 3$ Thus $\gamma(\Phi) \doteq .46$

(II) Newton's method

$$\Phi(x) = \tfrac{1}{2}(x - (bx + 2c)/(2x + b))$$

$p(\Phi) = 2$; $M(\Phi) = 1$

Thus $\gamma(\Phi) = 1$

(III) Halley's method

$$\Phi(x) \equiv x - ff'/(f'^2 - \tfrac{1}{2}ff'')$$

$$= (x-b)/3 + (b^2-4c)(2x+b)/3(3x^2+3bx + b^2 - c)$$

$p(\Phi) = 3$; $M(\Phi) = 2$

Thus $\gamma(\Phi) = .79$

Is it worth while to look further for a more efficient algorithm for quadratic equations? Surprisingly I can give a firm answer of "no" in the next section.

5. MAIN THEOREM

This is as good a place as any to show that γ is well-defined.

Lemma 1 For the algorithms and functions we are concerned with,

$$M(\Phi) > 0.$$

Proof If $M(\Phi) = 0$, then Φ is a linear function with rational coefficients and

$$\Phi(x) = x$$

is either identically true or has only a rational solution.

Q.E.D.

Liouville's theorem in number theory gives us:

Lemma 2 If α is an (irrational) algebraic number then there is a fixed m such that for all K:

$$|p/q - \alpha| < Kq^{-m}$$

for only finitely many pairs of integers p,q.

A stronger result which is not needed here is Roth's Theorem which replaces m by any number greater than 2 [Chapter XI, Hardy and Wright (1960)].

The Theorem If an (irrational) algebraic number α is approached by an iteration function Φ, which is rational with rational coefficients then:

$$\gamma(\Phi) \leqslant 1$$

Proof We can assume that $p(\Phi) > 1$ otherwise there is nothing to prove. Let x_o be a rational number which is close enough to α that, if $x_{n+1} = \Phi(x_n)$ for $n \geqslant 0$, $x_n \to \alpha$. For any $P < p(\Phi)$

$$|x_n - \alpha| < C^{P^n}$$

for some C such that 0<C<1 and n sufficiently large. But $x_n \equiv \Phi^{(n)}(x_o)$ is a rational, $x_n = p_n/q_n$ say, and so by Lemma 2,.

$$|p_n/q_n - \alpha| > Kq_n^{-m}$$

for some fixed m, K and all sufficiently large n. So $q_n^{\,m} > K.C^{-p^n}$ and therefore

(*) $q_n > D^{p^n}$ for some fixed D>1 and all sufficiently large n.

The next step is to show that to produce a rational with a large denominator needs a lot of M-operations. Given a finite set S of rationals, express them with a (positive) least common denominator Q as $\{p_1/Q, \ldots, p_s/Q\}$ and let

$$N_S = \max\{Q, p_1, \ldots, p_s\}$$

If $S_A = S \cup \{y\}$ where y is got from S by an A-operation, it can be checked that

$$N_{S_A} \leqslant k.N_S$$

where k depends only on the parameter λ (if any) of the operation. λ comes from a fixed finite set $\{\lambda_1, \ldots, \lambda_R\}$ associated with Φ.

The analogous result for M in place of A is:

$$N_{S_M} \leqslant (N_S)^2$$

Looking at the set of rationals $S^{(n)}$ generated from $S^{(o)} = \{x_o\}$ by n iterations of the algorithm Φ, we see that:

(**) $q_n \leqslant N_S(n) \leqslant E^{2^{n.M(\Phi)}}$ for some fixed E and all n.

We use here the fact that there is a bounded number of A- operations between successive M-operations.

Comparing (*) and (**) we get:

$$E^{2^{n.M(\Phi)}} > D^{P^n} \text{ for all sufficiently large n}$$

Hence,

$$M(\Phi) \geqslant \log_2 P$$

and since P was chosen arbitrarily, less than p(Φ),

$$M(\Phi) \geqslant \log_2 p(\Phi) \text{ and } \gamma(\Phi) \leqslant 1$$ Q.E.D.

Corollary Newton's method has optimal efficiency for quadratic equations.

It follows from my remarks in section 4 that the theorem holds also for iterations with memory. The method of the proof can also be used to show stronger results. For example, Φ may be allowed to grow a little more complicated with each iteration.

6. THE MOST EFFICIENT ALGORITHMS I CAN THINK OF ...

6.1 For high degree polynomials

If $\Phi_{e,m}$ is an inverse interpolatory iteration as defined by Traub (1964A), which uses $f, f', \ldots, f^{(e-1)}$ at $x_{i-1}, \ldots, x_{i-m}$ to compute x_i then

$$p(\Phi_{e,m}) \to e+1 \text{ as } m \to \infty$$

To compute $\{f(x), f'(x), \ldots, f^{(e-1)}(x)\}$ where f is a polynomial of large degree, I shall first compute $x^2, x^3, \ldots, x^k$, and $x^{2k}, x^{4k}, x^{8k}, \ldots$ until x^n, for a suitably chosen k. Then using the techniques described by Paterson and Stockmeyer (1971A), compute the e functions required in approximately en/2k further M-operations. The choice of $k = (\frac{1}{2}en)^{\frac{1}{2}}$ gives a total number of M-operations which is asymptoticly $(2en)^{\frac{1}{2}}$ as $n \to \infty$ for fixed e. The evaluation of $\Phi_{e,m}$ from $f, f', \ldots, f^{(e-1)}$ uses a number of operations which is independent of n, so

$$M(\Phi_{e,m}) \sim (2en)^{\frac{1}{2}} \text{ as } n \to \infty$$

For arbitrary $\delta > 0$, m can be chosen so that

$$p(\Phi_{e,m}) > e + 1 - \delta$$

So

$$\gamma(\Phi_{e,m}) > (2en)^{-\frac{1}{2}} \log(e + 1 - \delta)$$

for m,n sufficiently large. The integer which maximizes $e^{-\frac{1}{2}} \log(e + 1)$ is 4. m can be chosen so that $\gamma(\Phi_{4,m})$ is arbitrarily close to $(8n)^{-\frac{1}{2}} \log 5 \doteq .82\, n^{-\frac{1}{2}}$

I have not thought of any way to better this value and should be interested to hear of any improvement.

6.2 For $x^n - A = 0$ where n is large and A is rational

If $x^n - A = \delta$ for some small δ then

$$A^{1/n} = x(1 - \delta/x^n)^{1/n}$$

When the right hand side is expanded in powers of δ and truncated before the δ^p term, the result is an iteration function α_p of order p.

Replacing δ by $x^n - A$, α_p can be expressed in the form:

$$\alpha_p(x) = x \,.\, \sum_{i=o}^{p-1} a_i \,.\, (x^{-n})^i$$

If I first compute x^{-n} in log n + o(log n) M-operations [Section 4.6.3, Knuth (1969A)] then evaluate the polynomial in x^{-n} by the techniques of Paterson and Stockmeyer (1971A), α_p can be computed in log n + $(2p)^{\frac{1}{2}}$+o(log n) + O(log p) M-operations. With $(2p)^{\frac{1}{2}}$ = log n/log log n, I get

$$\gamma \sim 2 \log \log n/\log n$$

6.3 For $x^n - A = 0$ where n is small

In this case it is better to use a rational approximation to the power series obtained in (6.2). For example, there is an iteration function with order 13 for $x^5 - A = 0$ of the form:

$$x. \frac{a_o + a_1x^5 + a_2x^{10} + \ldots + a_6x^{30}}{b_o + b_1x^5 + b_2x^{10} + \ldots + b_6x^{30}}$$

If I compute x^5, x^{10}, x^{15} in 5 M-operations, and use only A-operations to compute

$$a_o + a_1x^5 + a_2x^{10} + a_3x^{15},$$

$$a_4x^5 + a_5x^{10} + a_6x^{15},$$

and similarly for the denominator, then only 9 M-operations are needed in total.

$$\gamma = (\log 13)/9 \doteq .41$$

which is the best I can manage.

A similar approach for $x^4 - A = 0$ works out best with a formula

$$x. \frac{a_o + a_1x^4 + a_2x^8 + a_3x^{12}}{b_o + b_1x^4 + b_2x^8 + b_3x^{12}}$$

of order 7 which requires at most 6 M-operations, and so $\gamma \doteq .47$ For this special case of $n = 4$ however, there is an indirect alternative which achieves efficiencies arbitrarily close to $\frac{1}{2}$. In effect we can compute $A^{\frac{1}{2}}$ and $A^{\frac{1}{4}}$ in parallel using the optimal square root algorithm.

Let $\psi(x,y) = \frac{1}{2}x + \frac{1}{2}y/x$. $M(\psi) = 1$ irrespective of whether y is a fixed parameter or not. Consider the following algorithm θ.

Given x (an approximant to $A^{\frac{1}{4}}$), compute
$y_o = x^2, y_1, \ldots, y_r$ where $y_{i+1} = \psi(y_i, A)$
for $i = 0, \ldots, r - 1$. Then compute
$z_o = x, z_1, \ldots, z_r$ where $z_{i+1} = \psi(z_i, y_r)$
for $i = 0, \ldots, r - 1$. $\theta(x) = z_r$.

If $x = A^{\frac{1}{4}} + \varepsilon$ then $y_r = A^{\frac{1}{2}} + O(\varepsilon^{2^r})$

Therefore $x = (y_r)^{\frac{1}{2}} + O(\varepsilon)$ and so

$$\theta(x) = z_r = (y_r)^{\frac{1}{2}} + O(\varepsilon^{2^r}) = A^{\frac{1}{4}} + O(\varepsilon^{2^r})$$

Thus $p(\theta) = 2^r$, $M(\theta) = 2r + 1$ and $\gamma(\theta) = r/2r + 1$

By choosing r sufficiently large, γ can be made arbitrarily close to $\frac{1}{2}$. This is the most efficient method I know to approach $A^{\frac{1}{4}}$. Using the same procedure we have:

Theorem If α can be approximated by an algorithm of efficiency γ then $\alpha^{\frac{1}{2}}$ can be approximated with an efficiency arbitrarily close to $\gamma/\gamma+1$.

For cubic equations either in general or in the special case $x^3 - A = 0$, I cannot improve on Newton's formula which for $ax^3 + bx^2 + c\,x + d = 0$ becomes

$$\Phi(x) = 2x/3 - (bx^2 + 2cx + 3d)/3(3ax^2 + 2bx + c)$$

$p(\Phi) = 2$, $M(\Phi) = 2$ and so $\gamma(\Phi) = \frac{1}{2}$

7. CONJECTURES AND CONCLUSION

Conjecture 1 The optimal efficiency for a general polynomial of degree n is $O(n^{-\frac{1}{2}})$

Conjecture 2 The optimal efficiency for polynomials of the form $x^n - A$ is $O(\log \log n/\log n)$

The only progress towards the establishment of these conjectures which I can report is the theorem by Paterson and Stockmeyer (1971A), proving that evaluating a general polynomial of degree n requires $O(n^{\frac{1}{2}})$ M-operations. A modest target would be to lower the upper bound of 1 in the main theorem of this paper to some function which tends to zero as n increases but I cannot suggest any ideas towards this goal.

"Conclusion" in the section title merely means "end".

PARALLEL ITERATION METHODS *

Shmuel Winograd

Mathematical Sciences Department
IBM Thomas J. Watson Research Center
Yorktown Heights, New York

I. INTRODUCTION

One of the problems which arise from the increasing use of multiprocessing is the efficient utilization of all the processors. If the solution of a problem requires T units of time when only one processor is used, it is hoped that using k processor will require only T/k units of time. Dorn (1962A) considered the problem of evaluating a polynomial, and showed that for k small compared with the degree n of the polynomial, a modification of Horner's method requires about n/k additions and n/k multiplications. Munro and Paterson (1971A) proved that for any k and n, a bound on the number of operations required to evaluate an n degree polynomial is about $2n/k + \log_2 k$, and showed a method which approaches this bound. Thus, for polynomial evaluation, one can achieve a good utilization of all the processes when their number is much smaller than the degree of the polynomial.

The problem of searching for the maximum of a unimodal function is completely different. Kiefer (1953A) showed that the best sequential method for determining an interval ε where the maximum of a unimodal function on $[0, 1]$ lies requires

* This work was supported (in part) by the Office of Naval Research under contract number N0014-69-C-0023.

about $-\log_\lambda \varepsilon$ iteration steps, where $\lambda = \frac{1+\sqrt{5}}{2}$. Karp and Miranker (1968A) investigated the same problem when k processors are available. They showed that in this case the number of iteration steps is about $-\log_{(k+2)/2} \varepsilon$. That means that if our problem is to determine, within ε, the position of the maximum of a unimodal function on $[0,1]$, and if it requires T units of time using one processor, it will require about $T/\log_\lambda k$ units of time when k processors are available.

In this paper, we will investigate the possible gain in speed for two kinds of iterative processes. We will consider local iterative schemes for finding the zero of an analytic function. The fastest rate of convergence for sequential iterative schemes was studied by Winograd and Wolfe (1971A). They showed that the power of convergence of an iterative scheme, which uses up to d derivatives, is at most d+2. That means that if it is required to know the zero with n digits accuracy, the number of iterations necessary grow as $\log_{d+2} n$. We will show that when m processors are available, the rate of convergence cannot exceed $1+(d+1)k$ and, therefore, that about $\log_{1+(d+1)k} n$ digits accuracy. Therefore, the gain in time grows only logarithmically with k.

II. ITERATION METHODS

Let f(x) be a function of a single real (or complex) variable, and assume that $f(r) = 0$. We will also assume that r is a simple zero. We will consider iterative scheme for finding r when we are given some points near r. In his book, J. Traub (1964A) describes various methods for finding r. If x_i denotes the ith approximation to r, and x_{n-j} $(1 \leq j \leq k)$ as well as $f(x_{n-j}), f'(x_{n-j}), \ldots, f^{(d)}(x_{n-j})$ $(1 \leq j \leq k)$ are known, we define P(x) as the minimum degree polynomial such that $P^{(i)}(x_{n-j}) = f^{(i)}(x_{n-j})$ $(0 \leq i \leq d,\ 1 \leq j \leq k)$, and choose x_n as the root of P closest to x_{n-1}. It is known (Traub, 1964A) $\lim_{n\to\infty} \frac{|x_n - r|}{|x_{n-1} - r|^\lambda} < \infty$ where λ is the largest root of the polynomial $z^k = (d+1) \sum_{i=0}^{k-1} z^i$, and that for all $\lambda' > \lambda$, we obtain

$\lim_{n\to\infty} \frac{|x_n - r|}{|x_{n-1} - r|^{\lambda'}} = \infty$. The number λ is called the power of convergence of the method. Thus, if x_{n-1} agrees with r to v significant digits, then x_n agrees with r to roughly λv digits. The number λ satisfies $d+1 \leq \lambda < d+2$, and approaches $d+2$ very rapidly as k increases. More precisely, Traub shows that $d+2 - \frac{e(d+1)}{(d+2)^k} \lambda < d+2 - \frac{d+1}{(d+2)^k}$

The same result holds even when we take other iterative methods. For example, if we define $g(y) = f^{-1}(y)$, (since $f'(r) \neq 0$, g exists at a neighborhood of r), then we define the polynomial $Q(y)$ as the minimal degree polynomial such that $Q^{(i)}(y_{n-j}) = g^{(i)}(y_{n-j})$ $(0 \leq i \leq d,\ 1 \leq j \leq k)$ where $y_{n-j} = f(x_{n-j})$. The next approximation x_n is defined by $x_n = Q(0)$. Using this method, we again obtain that the power of convergence is the largest root of the polynomial $z^k = (d+1) \sum_{i=0}^{k-1} z^i$.

In fact, it can be shown (Winograd and Wolfe, 1971A) that for every iterative method, which uses up to the dth derivative, there exists an analytic function f with a simple root r such that $\overline{\lim} \frac{|x_n - r|}{|x_{n-1} - r|^{d+2}} \geq 1$. That means that the power of convergence is at most $d+2$ no matter what iterative method is used; it cannot converge any faster. As we saw, we can achieve power of convergence of $d+2-\varepsilon$ by choosing k large enough.

III. PARALLEL ITERATION METHODS

The results mentioned above deal with sequential methods; that is, at every step, only one more approximating point is calculated. If we assume that we have m processors at our disposal, we can then compute, at each step, m new points $x_{n,1}, x_{n,2}, \ldots, x_{n,m}$. Of these points, we choose one of them and denote it by y_n. Of course, we will choose y_n as that $x_{n,i}$ which is closest to r, i.e., which approximates r best. Because of the assumption that $f'(x)$ does not vanish at the neighborhood of r, which includes all the approximations, we can determine y_n by examining $f(x_{n,j})$. If $x_{n,i}$ is the point which minimizes $|f(x_{n,j})|$, then we choose $y_n = x_{n,i}$. As usual, we define the error after the nth iteration as $e_n = |y_n - r|$. The question is how fast does e_n converge to 0.

Before stating the results, we need a few definitions.

Definition 1. An iteration scheme using d derivatives with parallelism m is a sequence $\{\varphi_k : k > k_0\}$, φ_k being a function of $(d+2)km$ real variables and range in the m_0-dimensional Euclidean space. We write φ_k as $\varphi_k(x_{k-j,i}, z_{k-j,i}, \ldots, z^d_{k-j,i})$ $(1 \leq j \leq k-1,\ 1 \leq i \leq m)$. (Note that φ_k is not necessarily defined everywhere.) Given such a scheme, the function $f \in C^d$, and the "starting values" $x_1, x_2, \ldots, x_{k_0}$, the associated iteration sequence is defined recursively by letting $z^n_{k-j,i} = f^{(n)}(x_{k-j,i})$, and $x_{k,i}$ as the ith component of φ_k. If φ_k is not defined, the sequence is terminated at $k-1$.

Definition 2. A local iteration scheme using d derivatives with parallelism m is a system consisting of $\{\varphi_k : k > k_0\}$ and a family $\{(I_t, J^0_t, J^1_t, \ldots, J^P_t)\}$ of sets of $p+1$ non-empty open intervals indexed by t, and having the properties:

a) If $(I_t, J^0_t, J^1_t, \ldots, J^P_t)$ is in the family and $I^*_t \subseteq I_t$, then so is $(I^*_t, J^0_t, J^1_t, \ldots, J^P_t)$.

b) $0 \notin J^1_t$.

The scheme converges for a function f whenever $x_{i,j}$ $(1 \leq i \leq k,\ 1 \leq j \leq m)$ satisfy $x_{i,j} \in (r-\Delta, r+\Delta) \subseteq I_t$, then:

1) For all i, j, $x_{i,j} \in I_t$ and $f^{(k)}(x_{i,j}) \in J^k_t$, $k = 0, 1, 2, \ldots, p$.

2) $\lim_{k\to\infty} \min_{1 \leq i \leq m} |x_{k,i} - r| = 0$.

That is when we start the method with approximations which are close enough to the root, then the whole iteration sequence remains close to the root and converges to it.

As an example, we will take the case $d = 0$, $m = 2$. The iteration scheme is defined by

$$x_{k,1} = x_{k-1,1} - \frac{x_{k-1,1} - x_{k-1,2}}{f(x_{k-1,1}) - f(x_{k-1,2})} f(x_{k-1,1})$$

$$x_{k,2} = x_{k,1} + (f(x_{k-1,1}))^5 .$$

It is easily verified that the power of convergence of this method is 2.

This example is really a variation of Newton's method in which we estimate the derivative $f'(x_{k-1,1})$ by $\frac{f(x_{k-1,1}) - f(x_{k-1,2})}{x_{k-1,1} - x_{k-1,2}}$. The secant method, which can also be viewed a variation of Newton's method, has a power of convergence of $\lambda_0 \approx 1.618$, and not 2, because of errors in the estimation of $f'(x_{k-1})$. In the method described above, the error in the estimation of $f'(x_{k-1})$ can be made so small as to not affect the power of convergence.

The same approach enables us, by remembering the use points, to obtain parallel iterative methods with $d = 0$, $m = 2$, whose power of convergence is as close to 3 as desired. Similarly, when $d = 0$, $m = 3$, we can obtain as good approximations of $f'(x_{k-1,1})$ and $f''(x_{k-1,1})$ as necessary, and therefore to obtain iterative methods whose power of convergence is $4-\varepsilon$.

Even when $d > 0$, we can generate parallel iterative schemes in a similar way. For example, if $d = 1$, $m = 2$, that is, when we know $f'(x_{k-1,1})$ and $f'(x_{k-1,2})$ in addition to $f(x_{k-1,1})$ and $f(x_{k-1,2})$, we can force $x_{k-1,1}$ and $x_{k-1,2}$ to be so close to each other to yield a good approximation of $f''(x_{k-1,1})$ and $f'''(x_{k-1,1})$, and thus using one of the known methods to obtain an iteration method whose power of convergence is very close to 5. In general, this method of generating parallel iteration methods yields powers of convergence arbitrarily close to $1+(d+1)m$. As we will see in the next section, this is about as fast a rate of convergence as can be achieved.

IV. THE UPPER BOUND

Theorem. For every local iteration scheme using d derivatives with parallelism m, there exists an analytic function f for which

$$\underline{\lim} \frac{e_k}{\prod_{j=1}^{k-1} e_r^{(d+1)m}} > 0$$

where $e_j = \min_{1 \leq i \leq m} |x_{j,i} - r|$.

The proof of this theorem is very similar to that of Theorem 2 in Winograd and Wolfe (1971A), so we will only sketch it.

Let $\{\varphi_k\}$ be an iteration scheme, and let f_0 be an analytic function with a simple root r_0. We apply the iteration scheme to f_0 until we obtain for the first time that $\dfrac{|x_{n,i} - r_0|}{\prod_{j=1}^{n-1}\prod_{i=1}^{m} |x_{j,i} - r_0|^{d+1}} < a$.

We then define $f_1 = f_0 + u \prod_{j=1}^{n-1}\prod_{i=1}^{m} (x - x_{j,i})^{d+1}$ and apply $\{\varphi_k\}$ to f_1, noting that in the first $n-1$ iterations we obtain the same points. The numbers a and u can be so chosen that

$$\frac{|x_{n,i} - r_1|}{\prod_{j=1}^{k-1}\prod_{i=1}^{m} |x_{j,i} - r_1|^{d+1}} > a \text{ for all } 1 \leq k \leq n,\ 1 \leq i \leq m, \text{ when } r_1$$

is the root of f_1 which is closest to r_0. We continue this way and obtain $f_2, f_3, \ldots$.

The desired function f is defined by $f(x) = \lim_{n\to\infty} f_n(x)$, for all points x in the interval in question. It is easily verified that the function f so defined is again an analytic function, being the uniform limit of analytic functions. Let r denote its root closest to r_0.

There is a number c (which depends on a and u) such that if we apply $\{\varphi_k\}$ to f, we obtain that for every i and k

$$\frac{|x_{k,i} - r|}{\prod_{j=1}^{k-1} \prod_{i=1}^{m} |x_{j,i} - r|^{d+1}} > c > 0.$$

Therefore, we obtain that $\dfrac{e_k}{\prod_{j=1}^{k-1} \prod_{i=1}^{m} |x_{j,i} - r|^{d+1}} > c > 0$ and

replacing $|x_{j,i} - r|$ by e_j, we obtain that $\underline{\lim}_{k \to \infty} \dfrac{e_k}{\prod_{j=1}^{n-1} e_j^{(d+1)m}} > c > 0.$

Corollary. For every local iteration scheme using d derivatives with paralelism m, there exists an analytic function f for which

$$\overline{\lim_{k \to \infty}} \frac{e_k}{e_{k-1}^{1+(d+1)m}} \geq 1.$$

Proof. Choose f as in the theorem and since

$$\frac{e_k}{\prod_{j=1}^{k-1} e_j^{(d+1)m}} = \prod_{r=2}^{k} \frac{e_j}{e_{j-1}^{1+(d+1)m}} \cdot e_1,$$ we obtain the results.

The corollary shows that no local iteration scheme using d derivatives with parallelism m can have power of convergence greater than $1+(d+1)m$. In the previous section, we showed that this bound can be approached.

V. CONCLUSIONS

In order to assess the gain obtained by parallelism, it is more convenient to reinterpret the notion of the power of convergence. Thus, if the power of convergence of an iteration scheme is λ and the root r is desired to accuracy ε, the number of iterations needed grows as $\log_\lambda \log 1/\varepsilon$. If we compare two iteration methods using d derivatives, one of which is sequential and the other has parallelism m, we see that the sequential method can reach accuracy ε with about $N = \log_{d+2} \log 1/\varepsilon$ iterations, the method using parallelism m requires at least $\log_{1+(d+1)m} \log 1/\varepsilon = N \cdot \log_{d+2}(1+(d+1)m)$. That u, for a fixed d, the number of iteration steps saved by the use of m-fold parallelism is proportional to $\log m$ and not to m. That is, the use of parallelism for this problem is "wasteful."

In reaching this conclusion, we should remember that we used the number of steps as our measure, and did not pay attention to the amount of work required in each step. It may be that the best way of using parallel capabilities for this type of problem is by using a sequential method, and utilizing the parallelism to speed up each step.

As was mentioned in the introduction, a logarithmic speed-up for an iterative method was also obtained by Karp and Miranker (1968A) when they considered an entirely different problem. The problem they considered can be reduced to the following combinatorial problem. In the interval $(0, 1)$ we choose m points, and thus divide it into $m+1$ intervals. Of these $m+1$ intervals, an opponent chooses two consecutive subintervals. We then add m points to the two intervals which were chosen and the opponent again chooses two consecutive subintervals. Our objective is to minimize the length of the subintervals chosen by the opponent, while his is to maximize it.

If we denote by $n(\varepsilon)$ the number of iterations needed to assure that the length of the chosen intervals does not exceed ε, then $n(\varepsilon)$ decreases only logarithmically with m. A. Lempel, jointly with the author, showed that this logarithmic gain holds even when the problem is generalized to the choice of t subintervals. This result, together with the one reported in the paper, suggest that it may be possible to define a wide class of iterative problems for which the gain resulted from parallelism is only logarithmic.

THE COMPUTATIONAL COMPLEXITY OF ITERATIVE METHODS FOR SYSTEMS OF NONLINEAR EQUATIONS

Richard Brent

Mathematical Sciences Department
IBM Thomas J. Watson Research Center
Yorktown Heights, New York

I. INTRODUCTION

Suppose that an iterative method M generates successive approximations $\underset{\sim}{x}_0, \underset{\sim}{x}_1, \ldots$ to a solution $\underset{\sim}{x}^*$ of the system

$$\underset{\sim}{f}(\underset{\sim}{x}) = \underset{\sim}{0} \tag{1.1}$$

of n nonlinear equations in n unknowns. If w_i is the amount of work required to compute $\underset{\sim}{x}_i$ from $\underset{\sim}{x}_{i-1}$ (and other results saved from previous iterations), then we say that the efficiency of M (for the given $\underset{\sim}{f}$, $\underset{\sim}{x}_0$ etc.) is

$$E = \lim_{i\to\infty} \frac{1}{w_{i+1}} \log \left(\frac{\log ||\underset{\sim}{x}_{i+1} - \underset{\sim}{x}^*||}{\log ||\underset{\sim}{x}_i - \underset{\sim}{x}^*||} \right), \tag{1.2}$$

if the limit exists. If a method M' produces successive approximations $\underset{\sim}{x}_i'$ with work w_i', then we say that M' has efficiency at least E if (1.2) holds for some w_i and $\underset{\sim}{x}_i$ satisfying $w_i' \le w_i$ and $||\underset{\sim}{x}_i' - \underset{\sim}{x}^*|| \le ||\underset{\sim}{x}_i - \underset{\sim}{x}^*||$.

Our aim is to compare the efficiencies of certain methods with the best possible, so we consider only methods with positive efficiency. For technical reasons, we assume that

$$0 < \liminf_{i\to\infty} w_i \le \limsup_{i\to\infty} w_i < \infty. \tag{1.3}$$

(Usually w_i is constant for sufficiently large i.) We assume that $\underset{\sim}{f}$ has continuous partial derivatives of all orders and that the Jacobian matrix of $\underset{\sim}{f}$ at $\underset{\sim}{x}^*$ is nonsingular. We say that the efficiency of M (<u>independent</u> of a particular choice of $\underset{\sim}{f}$, $\underset{\sim}{x}_0$ etc.) is E if E is the supremum of numbers E' such that M has efficiency at least E' for all functions $\underset{\sim}{f}$ (as above) and sufficiently good starting values.

The <u>order</u> of a method (for given $\underset{\sim}{f}$, $\underset{\sim}{x}_0$ etc.) is

$$\rho = \lim_{i\to\infty} \frac{\log \|\underset{\sim}{x}_{i+1} - \underset{\sim}{x}^*\|}{\log \|\underset{\sim}{x}_i - \underset{\sim}{x}^*\|}, \tag{1.4}$$

if the limit exists. The definitions of a method with order <u>at least</u> ρ, and with order ρ independent of a particular $\underset{\sim}{f}$, $\underset{\sim}{x}_0$ etc., are apparent.

The definition (1.2) has the following nice properties.

1. E is independent of the particular vector norm used (and similarly for ρ).

2. If ρ and $w = \lim_{i\to\infty} w_i$ exist, then $E = \frac{\log \rho}{w}$ is the logarithm of the "efficiency index" of Ostrowski (1960a). It follows from Gentleman (1971a) that any "reasonable" measure of computational efficiency is a monotonic function of E.

3. If methods M, M' have efficiencies E, E' and require $W(\varepsilon)$, $W'(\varepsilon)$ units of work to find an $\underset{\sim}{x}_i$ such that $\|\underset{\sim}{x}_i - \underset{\sim}{x}^*\| < \varepsilon$, then

$$\lim_{\varepsilon\to 0+} \frac{W(\varepsilon)}{W'(\varepsilon)} = \frac{E'}{E} \tag{1.5}$$

Thus, M requires E'/E times as much work as M' to reduce $\|\underset{\sim}{x}_i - \underset{\sim}{x}^*\|$ to a small positive tolerance.

Except for a brief comment in Section 5, we restrict our attention to methods which depend on the sequential evaluation of $\underset{\sim}{f}(\underset{\sim}{x})$ at certain points $\underset{\sim}{x}$, and the unit of computational work is one such evaluation. Thus, we neglect the possibility of evaluating derivatives of $\underset{\sim}{f}$ except by finite differences, and any overhead, i.e., work other than function evaluations, is ignored (except in Section 4).

2. MULTIVARIATE POLYNOMIAL INTERPOLATION METHODS

Suppose that $m \geq 1$, $n \geq 1$, $N = \binom{n+m}{m} = \frac{(n+m)!}{m!n!}$, and initial distinct approximations $\underset{\sim}{x}_0, \ldots, \underset{\sim}{x}_{N-1}$ are given. The inverse polynomial interpolation method $I_{m,n}$ generates $\underset{\sim}{x}_N, \underset{\sim}{x}_{N+1}, \ldots$ in the following way. Suppose that, for some $i \geq N$, approximations $\underset{\sim}{x}_0, \ldots, \underset{\sim}{x}_{i-1}$ have been generated. Then

$$\underset{\sim}{x}_i = \begin{pmatrix} a^{(1)} \\ \cdot \\ \cdot \\ \cdot \\ a^{(n)} \end{pmatrix}, \qquad (2.1)$$

where $a^{(1)}, \ldots, a^{(n)}$ are the constant terms in the multivariate polynomials

$$P_j(\underset{\sim}{y}) = a^{(j)} + \sum_{1 \leq k \leq n} b_k^{(j)} y_k + \ldots + \sum_{1 \leq k_1 \leq \ldots \leq k_m \leq n} c^{(j)}_{k_1, \ldots, k_m} y_{k_1} \ldots y_{k_m} \qquad (2.2)$$

which satisfy

$$\underset{\sim}{x}_p = \begin{pmatrix} P_1(\underset{\sim}{f}(\underset{\sim}{x}_p)) \\ \cdot \\ \cdot \\ \cdot \\ P_n(\underset{\sim}{f}(\underset{\sim}{x}_p)) \end{pmatrix} \qquad (2.3)$$

for $i-N \leq p < i$. (Solving the linear equations which give $\underset{\sim}{x}_i$ requires of order N^2 operations if a rank-one updating method is used.)

Let

$$\varepsilon_i = || \underset{\sim}{x}_i - \underset{\sim}{x}^* ||_2 . \qquad (2.4)$$

It is shown in Section 6 that, if $\varepsilon_{i-1}, \ldots, \varepsilon_{i-N}$ are sufficiently small, then

$$\varepsilon_i \le \frac{c}{|\Delta_i|} \prod_{j=0}^{m} \max\{\varepsilon_{i-1}, \ldots, \varepsilon_{i-\binom{n+j}{j}}\}, \tag{2.5}$$

where c is a constant (depending on $\underset{\sim}{f}$), and Δ_i is a certain N by N determinant (depending on $\underset{\sim}{f}(\underset{\sim}{x}_{i-1}), \ldots, \underset{\sim}{f}(\underset{\sim}{x}_{i-N})$) of order unity.

For the moment assume that

$$\limsup_{i \to \infty} |\log|\Delta_i||^{\frac{1}{i}} < \rho_{m,n}, \tag{2.6}$$

where $\rho_{m,n}$ is the (unique) positive real root of

$$\sum_{j=0}^{m} \rho_{m,n}^{-\binom{n+j}{j}} = 1. \tag{2.7}$$

If $\underset{\sim}{x}_i \to \underset{\sim}{x}^*$, then it follows from (2.5) and (2.6) that the order of convergence is at least $\rho_{m,n}$. (The proof is similar to some given in Brent (1972a).) Also, there are functions and starting points such that the order is exactly $\rho_{m,n}$. Hence, $I_{m,n}$ has efficiency

$$E_{m,n} = \log \rho_{m,n}. \tag{2.8}$$

If $1 < \rho < \rho_{m,n}$ and (Δ_i) is a sequence of independent random variables distributed so that $\sum_{i=1}^{\infty} P(|\Delta_i| \le \exp(-\rho^i))$ is convergent, then (2.6) holds with probability one. This suggests that, in some sense, the order of $I_{m,n}$ is "nearly always" at least $\rho_{m,n}$. However, the order may be less than $\rho_{m,n}$ if (2.6) does not hold (and the method breaks down if $\Delta_i = 0$).

$\rho_{m,n}$ and $E_{m,n}$ are monotonic increasing functions of m, so the efficiency of $I_{m,n}$ is bounded above by

$$E_{\infty,n} = \log \rho_{\infty,n}, \tag{2.9}$$

where $\rho_{\infty,n}$ is the (unique) real positive root of

$$\sum_{j=0}^{\infty} \rho_{\infty,n}^{-\binom{n+j}{j}} = 1. \tag{2.10}$$

In view of the results of Winograd and Wolfe (1971a) for n=1, the following conjecture is highly plausible.

Conjecture. No locally convergent method based entirely on function evaluations has efficiency greater than $E_{\infty,n}$.

It is easy to see that no method can have efficiency greater than log 2 : apply Winograd and Wolfe's result to a system of equations of the form

$$\left.\begin{aligned} f_1(x_1) &= 0 \\ &\vdots \\ f_n(x_n) &= 0 . \end{aligned}\right\} \qquad (2.11)$$

However,

$$E_{\infty,n} \sim \frac{\log n}{n} \qquad (2.12)$$

for large n, so our conjecture is much stronger than this.

Table 1 gives $E_{\infty,n}$ and $E_{1,n}/E_{\infty,n}$ for various values of n. Note that method $I_{1,n}$ has efficiency very close to $E_{\infty,n}$ if (2.6) holds. In fact,

$$1 - E_{1,n}/E_{\infty,n} = O(n^{-n/3}) \qquad (2.13)$$

as $n \to \infty$.

Table 1: Various Efficiencies

n	$k_S(n)$	$E_{\infty,n}$	$\frac{E_{1,n}}{E_{\infty,n}}$	$\frac{E_S(n)}{E_{\infty,n}}$
1	1	0.6931	0.6942	0.6942
2	3	0.4382	0.8724	0.6817
3	4	0.3414	0.9440	0.7048
4	4	0.2880	0.9763	0.7161
5	5	0.2532	0.9908	0.7227
10	8	0.1691	1.0000	0.7285
20	12	0.1084	1.0000	0.7417
50	23	0.0568	1.0000	0.7672
100	38	0.0337	1.0000	0.7874

3. SPECIAL CASES

Some special cases of the above results are of interest. If $n=1$, equation (2.10) reduces to

$$\sum_{j=0}^{\infty} \rho_{\infty,1}^{-(j+1)} = 1, \tag{3.1}$$

so $\rho_{\infty,1} = 2$ and $E_{\infty,1} = \log 2$. Thus, the result of Winograd and Wolfe (1971a) shows that the conjecture above is true for $n=1$.

If $n=m=1$, then (2.7) reduces to

$$\rho_{1,1}^{-1} + \rho_{1,1}^{-2} = 1, \tag{3.2}$$

so $\rho_{1,1} = (1+\sqrt{5})/2 = 1.618\ldots$, which is well known to be the order of the one-dimensional secant method (see Brent (1972a) or Ostrowski (1966a)). Rissanen (1971a) shows that, with certain restrictions, no method with the same memory can be more efficient.

If $n=1$, $m \geq 1$, then (2.7) reduces to

$$\sum_{j=0}^{m} \rho_{m,1}^{-(j+1)} = 1, \tag{3.3}$$

and $\rho_{m,1}$ is the order of the well-known (direct or inverse) m-th degree polynomial interpolation methods; see Traub (1964a).

If $n \geq 1$, $m=1$, then (2.7) reduces to

$$\rho_{1,n}^{-1} + \rho_{1,n}^{-(n+1)} = 1, \tag{3.4}$$

and $\rho_{1,n}$ is the order of Wolfe's secant method, provided (2.6) holds (this is much weaker than the assumption that Δ_i is bounded away from zero). See Wolfe (1959a), Barnes (1965a) and Bittner (1959a).

If $n=2$ then (2.10) reduces to

$$\phi(\rho_{\infty,2}^{-1}) = 2, \tag{3.5}$$

where $$\phi(x) = \sum_{j=0}^{\infty} x^{j(j+1)/2} . \qquad (3.6)$$

We note that

$$\phi(x) = \prod_{j=1}^{\infty} \left(\frac{1 - x^{2j}}{1 - x^{2j-1}} \right) \qquad (3.7)$$

by an identity of Gauss (see Hardy and Wright (1938a)). No generalization of (3.7) for $n > 2$ has been found.

4. PRACTICAL EFFICIENT METHODS

The methods $I_{m,n}$ of Section 2 are impractical if $N = \binom{m+n}{n}$ is large, for the overhead per function evaluation is of order N^2 operations. Also, their efficiency may be less than $E_{m,n}$ if (2.6) fails to hold. We shall briefly describe a class $\{ S_k \mid k \geq 1 \}$ of methods which avoid these disadvantages: the optimal method in the class has efficiency $E_S(n)$ close to $E_{\infty,n}$, and the overhead per function evaluation is of order n^2. (Since $\underset{\sim}{f}(\underset{\sim}{x})$ has n components $f_i(\underset{\sim}{x})$, each of which is a function of n variables, this is quite reasonable.)

If distinct approximations $\underset{\sim}{x}_i$ and $\underset{\sim}{x}'_i$ to a zero $\underset{\sim}{x}^*$ of $\underset{\sim}{f}(\underset{\sim}{x})$ have been found, then S_k generates approximations $\underset{\sim}{x}_{i+1}$ and $\underset{\sim}{x}'_{i+1}$ in the following way: if $\underset{\sim}{f}(\underset{\sim}{x}_i) = \underset{\sim}{0}$ then $\underset{\sim}{x}_{i+1} = \underset{\sim}{x}'_{i+1} = \underset{\sim}{x}_i$, otherwise do steps 1 to 4.

1. Let Q_i be an orthogonal matrix satisfying

$$\underset{\sim}{x}'_i = \underset{\sim}{x}_i + h_i Q_i \underset{\sim}{e}_1 , \qquad (4.1)$$

where $h_i = \| \underset{\sim}{x}_i - \underset{\sim}{x}'_i \|_2$ and

$$\underset{\sim}{e}_1 = (1, 0, \ldots, 0)^T .$$

2. Let A_i be the matrix whose j-th column is

$$A_i \underset{\sim}{e}_j = \frac{1}{h_i} [\underset{\sim}{f}(\underset{\sim}{x}_i + h_i Q_i \underset{\sim}{e}_j) - \underset{\sim}{f}(\underset{\sim}{x}_i)] .$$

3. Let $\underset{\sim}{y}_{i,0} = \underset{\sim}{x}_i$ and

$$\underset{\sim}{y}_{i,j} = \underset{\sim}{y}_{i,j-1} - J_i^{-1} f(\underset{\sim}{y}_{i,j-1}) \tag{4.3}$$

for $j = 1, \ldots, k$, where $J_i = A_i Q_i^T$.

4. Let $\underset{\sim}{x}_{i+1} = \underset{\sim}{y}_{i,k}$ and $\underset{\sim}{x}'_{i+1} = \underset{\sim}{y}_{i,k-1}$.

It is shown in Brent (1972b) that the efficiency of S_k is

$$E_S(k, n) = \frac{\log \frac{1}{2}(k + \sqrt{k^2 + 4})}{n + k - 1}. \tag{4.4}$$

If $k = k_S(n)$ is chosen so that $E_S(k, n)$ attains its maximum value $E_S(n)$, then

$$k_S(n) \sim n/\log n \tag{4.5}$$

and

$$E_S(n) \sim \frac{\log n}{n} \sim E_{\infty, n} \tag{4.6}$$

for large n. Table 1 gives $k_S(n)$ and $E_S(n)/E_{\infty,n}$ for various values of n. If the conjecture above is true, then the optimal method S_k is close to the best possible. In fact, we have

$$1 - \frac{E_S(n)}{E_{\infty, n}} = O\left(\frac{1}{\log n}\right) \tag{4.7}$$

as $n \to \infty$. It is an open question whether there are practical methods with efficiency lying between $E_S(n)$ and $E_{\infty,n}$.

5. METHODS WHICH USE COMPONENT EVALUATIONS

So far we have taken one evaluation of $\underset{\sim}{f}(\underset{\sim}{x}) = (f_1(\underset{\sim}{x}), \ldots, f_n(\underset{\sim}{x}))^T$ as the unit of computational work. If, instead, the evaluation of a component $f_i(\underset{\sim}{x})$ of $\underset{\sim}{f}(\underset{\sim}{x})$ is taken as $\frac{1}{n}$ units of work, then methods with efficiency greater than $E_{\infty,n}$ exist (at least for $n \geq 10$). In Brent (1972b) we describe a class $\{T_k \mid k \geq 1\}$ of methods related to Brown's method (see Brown and Conte (1967a), Rabin (1972a)). The optimal method in this class has efficiency

$$E_T(n) = \max_{k=1,2,\ldots} \frac{2\log(k+1)}{n+2k-1}, \tag{5.1}$$

and

$$E_T(n) \sim 2E_{\infty,n} \tag{5.2}$$

for large n. Whether significantly more efficient methods exist is an open question.

6. APPENDIX

In this appendix we sketch a proof of the inequality (2.5). Let $\underset{\sim}{g}$ be the inverse function of $\underset{\sim}{f}$, so

$$\underset{\sim}{g}(\underset{\sim}{f}(\underset{\sim}{x})) = \underset{\sim}{x} \tag{6.1}$$

for all $\underset{\sim}{x}$ sufficiently close to the simple zero $\underset{\sim}{x}^*$ of $\underset{\sim}{f}$. Let

$$\underset{\sim}{y}^{(p)} = \underset{\sim}{f}(\underset{\sim}{x}_p) \tag{6.2}$$

and

$$\eta_p = \| \underset{\sim}{y}^{(p)} \|_2 \tag{6.3}$$

for $p = i-1, \ldots, i-N$. By a renumbering, if necessary, there is no loss of generality in assuming that

$$\eta_{i-1} \le \eta_{i-2} \le \ldots \le \eta_{i-N}. \tag{6.4}$$

Let g_j be the j-th component of $\underset{\sim}{g}$, and $y_k^{(p)}$ the k-th component of $\underset{\sim}{y}^{(p)}$. By equations (2.2), (2.3), (6.1) and (6.2),

$$\begin{aligned} g_j(\underset{\sim}{y}^{(p)}) = a^{(j)} &+ \sum_{1 \le k \le n} b_k^{(j)} y_k^{(p)} + \ldots \\ &+ \sum_{1 \le k_1 \le \ldots \le k_m \le n} c^{(j)}_{k_1, \ldots, k_m} y^{(p)}_{k_1} \ldots y^{(p)}_{k_m} \end{aligned} \tag{6.5}$$

for $1 \le j \le n$ and $i-N \le p < i$.

Compare (6.5) with the Taylor series expansion

$$g_j(\underset{\sim}{y}) = A^{(j)} + \sum_{1\le k\le n} B_k^{(j)} y_k + \dots$$

$$+ \sum_{1\le k_1\le \dots \le k_m \le n} C^{(j)}_{k_1,\dots,k_m} y_{k_1} \dots y_{k_m}$$

$$+ R_j(\underset{\sim}{y}) \qquad (6.6)$$

of g_j about $\underset{\sim}{0}$. If $\alpha^{(j)} = a^{(j)} - A^{(j)}$ etc., then putting $\underset{\sim}{y} = \underset{\sim}{y}^{(p)}$ in (6.6) and subtracting (6.5) gives

$$\alpha^{(j)} + \sum_{1\le k\le n} \beta_k^{(j)} y_k^{(p)} + \dots + \sum_{1\le k_1\le \dots \le k_m \le n} \gamma^{(j)}_{k_1,\dots,k_m} y_{k_1}^{(p)} \dots y_{k_m}^{(p)} = R_j(\underset{\sim}{y}^{(p)}) \qquad (6.7)$$

for $1\le j\le n$ and $i-N\le p < i$. For each j, this gives a system of N linear equations in the N variables $\alpha^{(j)}, \beta_1^{(j)}, \dots, \gamma^{(j)}_{n,\dots,n}$. Solving by Cramer's rule for $\alpha^{(j)}$ gives

$$\alpha^{(j)} = D_1^{(j)} / D_2, \qquad (6.8)$$

where $D_1^{(j)}$ and D_2 are N by N determinants.

From the assumption (6.4) and the observation that $R_j(\underset{\sim}{y}^{(p)})$ is of order η_p^{m+1}, an inspection of the dominant terms in (6.8) shows that

$$|\alpha^{(j)}| \le \frac{K_j}{|\Delta_i|} \prod_{k=0}^{m} \eta_{i-\binom{n+k}{k}}, \qquad (6.9)$$

where K_j is a constant, and

$$\Delta_i = D_2 \prod_{k=1}^{m} \prod_{j=1+\binom{n+k-1}{k-1}}^{\binom{n+k}{k}} \eta_{i-j}^{-k} \qquad (6.10)$$

is of order unity.

From (6.1) and (6.6), it is immediate that the zero $\underset{\sim}{x}^*$ of $\underset{\sim}{f}$ is given by

$$\underset{\sim}{x}^* = \begin{pmatrix} A^{(1)} \\ \vdots \\ A^{(n)} \end{pmatrix} \quad . \tag{6.11}$$

Thus, from (2.1) and (6.9), we have

$$\| \underset{\sim}{x}_i - \underset{\sim}{x}^* \|_2 \le \frac{K}{|\Delta_i|} \prod_{k=0}^{m} \eta_{i-(n_k^+ k)} , \tag{6.12}$$

where

$$K = \left(\sum_{j=1}^{n} K_j^2 \right)^{\frac{1}{2}} . \tag{6.13}$$

In view of the assumption (6.4) and the fact that $\underset{\sim}{x}_{i-1}, \ldots, \underset{\sim}{x}_{i-N}$ are close to the simple zero $\underset{\sim}{x}^*$, the result (2.5) follows from (6.12).

THE COMPUTATIONAL COMPLEXITY OF ELLIPTIC

PARTIAL DIFFERENTIAL EQUATIONS*

Martin H. Schultz

Department of Computer Science

Yale University

1. INTRODUCTION

In this paper, we consider the computational complexity of the class of all procedures for computing a second order accurate approximation (on a square grid) to the solution of a linear, second order elliptic partial differential equation in a square domain. In particular, we present and analyze a new asymptotically optimal procedure for this problem. Moreover, we will show that a preconditioned form of our procedure is well-conditioned.

To be precise, we wish to approximate the function $u(x,y)$, $(x,y) \in S \equiv \{(x,y) \mid 0 < x, y < 1\}$ such that

$$(1) \quad -\frac{\partial}{\partial x}\left[p(x,y)\frac{\partial u}{\partial x}(x,y)\right] - \frac{\partial}{\partial y}\left[q(x,y)\frac{\partial u}{\partial y}(x,y)\right] + r(x,y)\,u(x,y)$$

$= f(x,y)$, for all $(x,y) \in S$, and

(2) $u(x,y) = 0$, for all $(x,y) \in \partial S \equiv$ the boundary of S, where

$p(x,y)$, $q(x,y)$, $r(x,y)$, and $f(x,y)$ are given smooth functions such that there exist two positive constants λ and γ satisfying

(3) $\lambda \leq p(x,y) \leq \gamma$, $\lambda \leq q(x,y) \leq \gamma$, and $\lambda \leq r(x,y) \leq \gamma$.

For every positive integer, N, we define $h \equiv (N+1)^{-1}$ and a square mesh $\mu(N)$: $\{(x(i), y(j)) \mid 0 \leq i, j \leq N+1\}$, where $x(i) \equiv i/(N+1)$ and $y(j) \equiv j/(N+1)$, $0 \leq i, j \leq N+1$. We consider the class of procedures, C, such that for each positive integer N, each procedure

*This research was supported in part by the Office of Naval Research, N0014-67-A-0097-0016.

yields an N×N array, $\bar{v} \equiv \{v(i,j) \mid 1 \leq i, j \leq N\}$, such that $\bar{v}$ is a second order approximation to $\bar{u} \equiv \{u(x(i), y(j)) \mid 1 \leq i, j \leq N\}$, i.e.,

$$(4) \quad \|\bar{u}-\bar{v}\|_2 \equiv \left(h^2 \sum_{1 \leq i,j \leq N} |u(x(i),y(j)) - v(i,j)|^2\right)^{1/2} = O(N^{-2})$$

as $N \to \infty$. In other words, our class of feasible procedures is exactly those which yield a "second order accurate" approximation to u(x,y) at the mesh points of $\mu(N)$. Our complexity criteria will be the number of arithmetic operations required to set up and solve an approximate finite dimensional problem and obtain $\bar{v}$ (assuming the functions p(x,y), q(x,y), r(x,y), and f(x,y) to be available only at the mesh points of $\mu(N)$) and the storage required.

Clearly we cannot have any procedure which requires fewer than $O(N^2)$ arithmetic operations and $O(N^2)$ storage locations as $N \to \infty$. In fact, even if we know the solution and only wish to tabulate it at the N^2 mesh points of $\mu(N)$, $\{(x(i),y(j)) \mid 1 \leq i, j \leq N\}$, we would require at least N^2 arithmetic operations and N^2 storage locations.

The best known finite difference procedures for this type of problem are all based on the five point finite difference approximation of the left-hand side of (1) coupled with an appropriate finite linear combination of point evaluations of the right-hand side of (1). These combinations lead to procedures which belong to C, our class of feasible procedures. Moreover, they all lead to linear systems of order N^2 with matrices of the form

$$h^{-2}\begin{bmatrix} 4 & -1 & & -1 & & 0 \\ -1 & \ddots & \ddots & & \ddots & \\ & \ddots & \ddots & \ddots & & -1 \\ -1 & & \ddots & \ddots & \ddots & \\ & \ddots & & \ddots & \ddots & -1 \\ 0 & & -1 & & -1 & 4 \end{bmatrix}$$

with band width N. These systems may be solved by a variety of excellent procedures from numerical linear algebra. For example, the band-Cholesky decomposition procedure requires $O(N^4)$ arithmetic operations and the successive overrelaxation iterative procedure requires $O(N^3 \ln_2 N)$ arithmetic operations to reduce the initial error by a factor of N^{-2} [Dorr (1970A), George (1971A), Varga (1962A)]. In this paper, we present and analyze a finite element procedure belonging to C which requires $O(N^2)$ arithmetic operations.

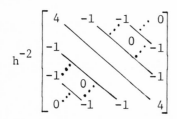

2. AN ASYMPTOTICALLY OPTIMAL PROCEDURE

In this section, we define a finite element procedure using piecewise bivariate cubic Hermite polynomials [Schultz (1973A)], defined on a "square root mesh", and special singular functions. This combination of basis functions has been extensively used by G. Fix (1968A) and others [Fix, Gulati and Wakoff (1972A)]. We will show that our procedure is in our feasible class and that its complexity is asymptotically the same as the trivial lower bounds given in the Introduction. Hence, our procedure will be shown to be asymptotically optimal.

To define our basis functions, let $n \equiv \lfloor N^{1/2} \rfloor$, i.e., the greatest integer less than or equal to $N^{1/2}$, $H \equiv (n+1)^{-1}$, $\Delta(n): 0 < H < 2H < \ldots < (n+1)H = 1$ be the uniform mesh on the interval [0,1], and $\rho(n) \equiv \Delta(n) \times \Delta(n)$ be the corresponding uniform square mesh on the unit square.

For each i with $1 \leq i \leq n$, we define

$$h_i(x) \equiv \begin{cases} -2H^{-3}[x-(i-1)H]^3 + 3H^{-2}[x-(i-1)H]^2, & \text{if } (i-1)H \leq x \leq iH, \\ 2H^{-3}[x-iH]^3 - 3H^{-2}[x-iH]^2 + 1, & \text{if } iH \leq x \leq (i+1)H, \text{ and} \\ 0 & \text{, if } x \notin [(i-1)H, (i+1)H] \end{cases}$$

and for each i with $0 \leq i \leq n+1$, we define

$$h_i^1(x) \equiv \begin{cases} H^{-3}[x-(i-1)H]^2\,[x-iH], & \text{if } (i-1)H \leq x \leq iH, \\ H^{-3}[x-iH]\,[(i+1)H-x]^2, & \text{if } iH \leq x \leq (i+1)H, \text{ and} \\ 0 & \text{, if } x \notin [(i-1)H, (i+1)H]. \end{cases}$$

Similarly, we define $h_j(y)$ for each $1 \leq j \leq n$ and $h_j^1(y)$ for each $0 \leq j \leq n+1$. Our piecewise bivariate cubic Hermite basis functions are defined to be

$$\{h_i(x)h_j(y),\ h_i(x)h_k^1(y),\ h_\ell^1(x)h_j(y),\ h_\ell^1 h_k^1(y) \mid 1 \leq i, j \leq n$$

and $0 \leq k, \ell \leq n+1\}$.

There are $4(n+1)^2$ such basis functions. Each interior mesh point (iH,jH), $1 \leq i, j \leq n$, has the four basis functions $h_i(x)h_j(y)$, $h_i^1(x)h_j(y)$, $h_i(x)h_j^1(y)$, and $h_i^1(x)h_j^1(y)$ associated with it, each corner mesh point has one basis function associated with it, and each of the other boundary mesh points has two basis functions associated with it.

The basic idea of our procedure is to use a fourth order accurate procedure on the square root mesh to yield a second order accurate procedure on the original mesh. However, this idea depends on the smoothness of the third and fourth derivatives of the solution u(x,y) which in these problems are generally singular. To alleviate this difficulty, we introduce three additional basis functions in the neighborhood of each corner of S. The asymptotic behavior of the third and fourth derivatives of these functions is the same as the asymptotic behavior of the corresponding derivatives of the solution [Lehman (1959A)]. For the corner at the origin (0,0) these are given in polar coordinates as follows:

$$S^1_{0,0}(r,\theta) \equiv \begin{cases} r^2(\ln r \sin 2\theta + \theta \cos 2\theta + \frac{\pi}{2}\sin^2\theta), \\ \qquad \text{if } 0 \le \theta \le \pi/2,\ 0 \le r \le 1/4, \\ p(r)\ (\ln r \sin 2\theta + \theta \cos 2\theta + \frac{\pi}{2}\sin^2\theta), \\ \qquad \text{if } 0 \le \theta \le \pi/2,\ 1/4 \le r \le 1/3, \\ 0, \qquad \text{otherwise}, \end{cases}$$

where p(r) is the unique seventh degree polynomial in r such that

$$p(1/4) = 1/16,\ \frac{dp}{dr}(1/4) = 1/2, \text{ and } \frac{d^tp}{dr^t}(1/4) = \frac{d^kp}{dr^k}(1/3) = 0$$

for $2 \le t \le 3$, $0 \le k \le 3$, $S^2_{0,0}(r,\theta) \equiv rS^1_{0,0}(r,\theta)\cos\theta$, and $S^3_{0,0}(r,\theta) \equiv rS^1_{0,0}(r,\theta)\sin\theta$. For the other three corners, the singular basis functions are analogous.

Thus, all together we have $4(n+1)^2 + 12$ basis functions. We order our basis functions by using the "natural" ordering of the piecewise bivariate cubic Hermite basis functions followed by any ordering of the singular basis functions, i.e., we consecutively order the basis functions associated with each mesh point of $\rho(n)$ and consecutively order the mesh points along rows.

Thus, we may rename our basis functions as $\{B_i(x,y) \mid 1 \le i \le M \equiv 4(n+1)^2 + 12\}$ and we seek an approximate solution of the form

$$w(x,y) \equiv \sum_{i=1}^{M} \beta_i B_i(x,y).$$

We determine the vector of coefficients $\underline{\beta} \in R^M$ by means of the Rayleigh-Ritz-Galerkin procedure [Schultz (1973A)]. That is, we seek those $\underline{\beta}$ such that

$$(5)\quad \int_0^1 \int_0^1 \left\{ p(x,y) \frac{\partial w}{\partial x} \frac{\partial B_j}{dx} + q(x,y) \frac{\partial w}{\partial y} \frac{\partial B_j}{\partial y} + r(x,y)wB_j \right\} dxdy$$

$$= \int_0^1 \int_0^1 f(x,y)B_j dxdy$$

for all $1 \leq j \leq M$.

This leads to a characterization of $\underline{\beta}$ as the solution of the M×M linear system

$$(6)\quad A\underline{\beta} = \underline{k}, \quad \text{where}$$

$$A \equiv [a_{ij}]$$

$$\equiv \left[\int_0^1 \int_0^1 \left\{ p(x,y) \frac{\partial B_i}{\partial x} \frac{\partial B_j}{\partial x} + q(x,y) \frac{\partial B_i}{\partial y} \frac{\partial B_j}{\partial y} + r(x,y)B_i B_j \right\} dxdy \right]$$

and $\underline{k} \equiv [k_i] \equiv \left[\int_0^1 \int_0^1 f(x,y)B_i dxdy \right]$. The matrix A of this system is symmetric, positive definite and hence the linear system has a unique solution $\underline{\beta}^+$. Moreover, the matrix A is sparse with the following regular structure:

$$A = \begin{bmatrix} A_{11} & A_{12} \\ A_{21}^T & A_{22} \end{bmatrix}, \text{ where } A_{11} \text{ is a } 4(n+1)^2 \times 4(n+1)^2 \text{ band matrix}$$

with band width $O(n)$, A_{12} is a $4(n+1)^2 \times 12$ matrix and A_{22} is a 12×12 matrix.

Since the coefficients of the differential equation are variable, the entries of A and $\underline{k}$ must be approximated numerically. For each $0 \leq i \leq N+1$, we define the "hat function"

$$L_i(x) \equiv \begin{cases} h^{-1}(x - (i-1)h), & \text{if} \quad (i-1)h \leq x \leq ih, \\ -h^{-1}(x - (i+1)h), & \text{if} \quad ih \leq x \leq (i+1)h, \\ 0 \quad , & \text{if } x \notin [(i-1)h, (i+1)h]. \end{cases}$$

Similarly, we define $L_j(y)$ for each $0 \leq j \leq N+1$. In (6) we approximate A by

$$\tilde{A} \equiv [\tilde{a}_{ij}]$$

$$\equiv \left[\int_0^1 \int_0^1 \left\{ \tilde{p}(x,y) \frac{\partial B_i}{\partial x} \frac{\partial B_j}{\partial x} + \tilde{q}(x,y) \frac{\partial B_i}{\partial y} \frac{\partial B_j}{\partial y} + \tilde{r}(x,y) B_i B_j \right\} dxdy \right]$$

and $\underline{k}$ by $\tilde{\underline{k}} \equiv [\tilde{k}_i] \equiv \left[\int_0^1 \int_0^1 \tilde{f}(x,y) B_i dxdy \right]$, where

$\tilde{p}(x,y) \equiv \sum_{0 \le i,j \le N+1} p(ih,jh) L_i(x) L_j(y)$, $\tilde{q}(x,y) \equiv \sum_{0 \le i,j \le N+1} q(ih,jh)$ $L_i(x) L_j(y)$, $\tilde{r}(x,y) \equiv \sum_{0 \le i,j \le N+1} r(ih,jh) L_i(x) L_j(y)$, and

$\tilde{f}(x,y) \equiv \sum_{0 \le i,j \le N+1} f(ih,jh) L_i(x) L_j(y)$ are the piecewise bilinear interpolates of $p(x,y)$, $q(x,y)$, $r(x,y)$, and $f(x,y)$ respectively. This yields the approximate linear system

(7) $\tilde{A}\underline{\beta}^* = \tilde{k}$,

whose unique solution $\underline{\beta}^*$ is used to compute the approximation

$$v(k,\ell) \equiv \sum_{i=1}^{M} \beta_i^* B_i(kh,\ell h), \quad 1 \le k,\ell \le N.$$

We can show that this procedure is well-defined and is in our class of feasible procedures. For all nonnegative integers p and all real numbers $1 \le q \le \infty$, we define the Sobolev norm

$$\|w\|_{p,q} \equiv \left[\int_0^1 \int_0^1 \sum_{j=0}^{p} \sum_{t=0}^{j} \left| \frac{\partial^j w}{\partial x^{j-t} \partial y^t} (x,y) \right|^q dxdy \right]^{1/q}$$

for all sufficiently smooth functions w defined on S. Moreover, we let $W^{p,q}$ denote the completion, with respect to the norm $\|\cdot\|_{p,q}$, of the real-valued infinitely differentiable functions on the unit square.

We will assume in the remainder of this paper that p,q, and $r \in W^{2,\infty}$, $f \in W^{2,2}$, and that f is analytic in a neighborhood of each corner. Using results of Lehman (1959A) and Schultz (1971A), we can prove the following theorem.

Theorem 1. The matrix $\tilde{A}$ is symmetric, positive definite, the linear system (7) has a unique solution $\underline{\beta}^*$, and our procedure is second order accurate, i.e., if $e(x,y) \equiv u(x,y) - \sum_{i=1}^{M} \beta_i{}^* B_i(x,y)$,

then $\left(\sum_{1 \leq i,j \leq N} h^2 e^2(ih,jh) \right)^{1/2} = O(N^{-2})$ as $N \to \infty$.

We can bound the number of arithmetic operations needed to compute $\tilde{A}$ and $\underline{\tilde{k}}$.

Theorem 2. The number of arithmetic operations required to compute $\tilde{A}$ and $\underline{\tilde{k}}$ is $O(N^2)$ as $N \to \infty$.

Proof. The matrix $\tilde{A}$ and the vector $\underline{\tilde{k}}$ have $O(N)$ nonzero entries each of which is obtained by formal integration using the piecewise, bilinear interpolates with respect to $\mu(N)$ of the coefficients of the differential equation in place of the coefficients. Since the first $4(n+1)^2$ basis functions are local with respect to $\rho(n)$, each of the nonzero entries of the submatrices $\tilde{A}_{11}$ and $\tilde{A}_{12}$ and each of the first $4(n+1)^2$ components of $\underline{\tilde{k}}$ can be computed with $O(N)$ arithmetic operations.

Since the last 12 basis functions are global, each of the 144 entries of $\tilde{A}_{22}$ and each of the last 12 components of $\underline{\tilde{k}}$ can be computed with $O(N^2)$ arithmetic operations. Hence, the total number of arithmetic operations to compute $\tilde{A}$ and $\underline{\tilde{k}}$ is $O(N^2)$. QED.

Our linear system (7) can be efficiently solved by the profile Cholesky decomposition procedure [George (1971A)]. That is, we compute a symmetric triangular decomposition of $A \equiv LL^T$ as follows. For each $1 \leq i \leq M$, let f_i denote the smallest nonnegative integer such that $a_{ij} = 0$ if $i-j > f_i$ and compute the lower triangular matrix L row by row by means of the formulae

(8) $\ell_{ij} \equiv 0, \quad \text{if } j \leq 0,\ j > i, \text{ or } i-j > f_i$

(9) $\ell_{ij} \equiv (a_{ij} - \sum_{k=i-f_i}^{j-1} \ell_{ik}\ell_{jk})/\ell_{jj}, \quad \text{if } i-f_i \leq j \leq i-1,$

and

(10) $\ell_{ii} \equiv (a_{ii} - \sum_{k=i-f_i}^{i-1} \ell_{ik}^2)^{1/2}$.

Having computed L, we then solve for $\underline{\beta}^* = (L^T)^{-1} L^{-1}\underline{\tilde{k}}$. Using a general result of George (1971A) on the complexity of the profile Cholesky decomposition, we may prove the following result.

Theorem 3. The computation of the solution of (7) by the profile Cholesky procedure requires $O(N^2)$ arithmetic operations and $O(N^2)$

storage locations as $N \to \infty$.

Finally, we obtain the approximate solution $\overline{v}$ by evaluating the expansion $\sum_{i=1}^{M} \beta_i^* B_i(x,y)$ at the N^2 mesh points of $\mu(N)$. That is, $v(k,\ell) \equiv \sum_{i=1}^{M} \beta_i^* B_i(kh,\ell h)$, $1 \leq k, \ell \leq n$. Since the evaluation of $\sum_{i=1}^{M} \beta_i^* B_i(x,y)$ at any point of S involves only nineteen of the basis functions, we may prove the last result.

Theorem 4. The procedure to evaluate the expansion $\sum_{i=1}^{M} \beta_i^* B_i(x,y)$ at the N^2 mesh points $\{(ih,jh) \mid 1 \leq i, j \leq N\}$ requires $0(N^2)$ arithmetic operations as $N \to \infty$.

3. A WELL-CONDITIONED ASYMPTOTICALLY OPTIMAL PROCEDURE

As is well known the behavior of the round-off error in elimination schemes for linear systems is related to the condition number of the associated matrix. We would expect that the condition number of the matrix $\tilde{A}$ of (7) to be large, because of the fact the last 12 rows are almost linearly dependent on the first $4(n+1)^2$ rows. In fact, using results of deBoor and Fix (1972A) we have

$$(11) \quad \inf\left\{ \left\| S^1_{i,j} - \sum_{t=1}^{4(n+1)^2} \alpha_t B_t \right\|_{k,2} \,\middle|\, \underline{\alpha} \in R^{4(n+1)^2} \right\} = 0\left(n^{-\frac{5}{2}+k}\right)$$

for $k = 0,1$ and $0 \leq i, j \leq 1$, and

$$(12) \quad \inf\left\{ \left\| S^\ell_{i,j} - \sum_{t=1}^{4(n+1)^2} \alpha_t B_t \right\|_{k,2} \,\middle|\, \underline{\alpha} \in R^{4(n+1)^2} \right\} = 0\left(n^{-\frac{7}{2}+k}\right)$$

for $\ell = 2,3$, $k = 0,1$, and $0 \leq i, j \leq 1$. However, numerical experiments of Fix and others indicate that in practice this ill-conditioning is not a problem. In this section, we will show that with preconditioning of the singular basis function, we can define a well-conditioned procedure which is an analogue of section 2.

We proceed basically in the same fashion as we did in

section 2 except that we now consider only those values of $N \equiv n^{3/2} \equiv (2^p-1)^{3/2}$ for some positive integer p, and define the piecewise, bivariate Hermite polynomials relative to these values of n. Moreover, we add in only the four singular basis functions $S^1_{i,j}(r,\theta)$, $0 \le i, j \le 1$ (one for each corner).

We require preconditioning of these four singular basis functions. This preconditioning depends on n, but not the differential equation. Hence, for each value of n we need do the preconditioning only once. Moreover, we will show that the work and storage requirements for this preconditioning are $O(N^2)$ as $N \to \infty$. It is not clear to us whether or not this preconditioning is intrinsic to the result or to the method of proof.

For $0 \le i, j \le 1$, we define $\tilde{S}^n_{i,j}(r,\theta)$ to be n^2 times the $L^2(S)$-orthogonal complement of $S^1_{ij}(r,\theta)$ with respect to the space spanned by the piecewise, bivariate Hermite polynomials with respect to $\rho(n)$. That is,

$$\tilde{S}^n_{i,j} \equiv n^2\left[S^1_{i,j} - \sum_{k=1}^{4(n+1)^2} \alpha_k B_k\right], \quad \text{where}$$

$$B\underline{\alpha} = \underline{g},$$

$$B \equiv [b_{\ell k}] \equiv \left[\int_0^1 \int_0^1 B_\ell(x,y)B_k(x,y)dxdy\right] \quad \text{and}$$

$$\underline{g} \equiv [g_k] \equiv \left[\int_0^1 \int_0^1 S^1_{i,j}\, B_k(x,y)dxdy\right]. \quad \text{[Schultz (1973A)]}.$$

As was shown by Schultz (1973A), the condition number of the matrix B is bounded independent of n and, using a special "nested ordering" of the mesh points of $\rho(n)$ (instead of the natural ordering) [George (1972A)], the system $B\underline{\alpha} = \underline{g}$ can be solved with a sparse Cholesky decomposition procedure in $O(n^3) = O(N^2)$ arithmetic operations. We may define George's ordering recursively. If we denote the mesh points of $\rho(n)$ by $\{(s(i), t(j)) \mid 1 \le i, j \le 2^p+1\}$, then we first order the set of mesh points of the form $(s(2k), t(2\ell))$, $1 \le k, \ell \le 2^{p-1}$, arbitrarily and remove them from the mesh $\rho(n)$. We are then left with a mesh of the same form as $\rho(n)$ except with p replaced by p-1. We continue this ordering procedure recursively as far as possible and then the remaining mesh points arbitrarily.

We now define $B_{4(n+1)+i+2j}(x,y) \equiv \tilde{S}^n_{i,j}$ for $0 \le i, j \le 1$, and we have a total of $M \equiv 4(n+1)^2+4$ basis functions. Proceeding as in section 2, we obtain an approximate solution

$\sum_{i=1}^{M} \beta_i^* B_i$ by solving the approximate Rayleigh-Ritz-Galerkin

equations

(13) $\tilde{A}\underline{\beta}^* = \underline{k}$, where $\tilde{A}$ and $\underline{k}$ are the analogue of the corresponding matrix and vector of section 2.

We use George's (1972A) "nested ordering" followed by the profile Cholesky decomposition procedure to solve the linear system (13) and obtain the approximate solution $\overline{v}$ by evaluating the expansion $\sum_{i=1}^{M} \beta_i^* B_i(x,y)$ at the N^2 mesh points of $\mu(N)$. We may verify that analogues of Theorems 1 - 4 hold for this procedure. Moreover, we can prove the following bound for the condition number of $\tilde{A}$.

Theorem 5. Condition of $\tilde{A} \equiv$ (maximum eigenvalue of $\tilde{A}$)/(minimum eigenvalue of $\tilde{A}$) $= O(N^2)$ as $N \to \infty$.

Proof. If $\tau^2 \equiv \underline{\beta}^T\tilde{A}\underline{\beta}$, it suffices to obtain upper and lower bounds for τ in terms of N. From inequalities (3) and (11), the results of Schultz (1973A), and the definition of the preconditioned singular basis functions, we have

$$\tau \le O\left(\left\| \sum_{i=1}^{4(n+1)^2} \beta_i B_i \right\|_{1,2}\right) + O\left(\left\| \sum_{i=4(n+1)^2+1}^{M} \beta_i B_i \right\|_{1,2}\right) \tag{14}$$

$$\le O\left(n^{1/2}\left(\sum_{i=1}^{4(n+1)^2} \beta_i^2\right)^{1/2}\right) + O\left(n^{1/2}\left(\sum_{i=4(n+1)^2+1}^{M} \beta_i^2\right)^{1/2}\right).$$

However, since $S_{i,j}^1 \notin W^{3,p}$ for any $p \ge 2$, we have from inequality (3), the results of Munteanu and Schumaker (1972A) giving "inverse theorems" for piecewise polynomial approximation, the results of Schultz (1973A), and the definition of the preconditioned singular basis functions, that there exists a positive constant K such that

$$\tau \ge K\left[O\left(n^{-1/2}\left(\sum_{i=1}^{4(n+1)^2} \beta_i^2\right)^{1/2}\right) + O\left(n^{-1}\left(\sum_{i=4(n+1)^2+1}^{M} \beta_i^2\right)^{1/2}\right)\right]. \tag{15}$$

Combining the inequalities (14) and (15), we have the bound cond $\tilde{A} = O(n^3) = O(N^2)$. QED.

If we broaden our class of feasible procedures slightly to include all those procedures for which the error at the mesh points of $\mu(N)$ behaves like $O(N^{-2+\varepsilon})$ for all $\varepsilon > 0$, as $N \to \infty$, then we can proceed as in this section except for leaving out the four singular basis functions.

The $4(n+1)^2 \times 4(n+1)^2$-matrix af this procedure is $\tilde{A}_{11}$ whose condition number is $O(N^2)$ as $N \to \infty$ and the procedure is asymptotically optimal. Moreover, this procedure is the simplest to program and most likely is the most efficient for practical problems [Eisenstat and Schultz (1972A)].

4. CONCLUSIONS

We have shown that our finite element procedure is second order accurate and requires $O(N^2)$ arithmetic operations and $O(N^2)$ storage locations as $N \to \infty$. Hence, it is asymptotically optimal and asymptotically at least as efficient as any other second order procedure. Indeed solving a second order, linear elliptic partial differential equation in a square is asymptotically no less efficient than tabulating the solution! Moreover, a preconditioned form of our procedure involves a matrix whose condition number is $O(N^2)$ as $N \to \infty$. For numerical experiments and further theoretical details see Eisenstat and Schultz (1972A).

ACKNOWLEDGEMENT

The author is grateful to S. C. Eisenstat for many helpful suggestions regarding the content of this paper.

REDUCIBILITY AMONG COMBINATORIAL PROBLEMS†

Richard M. Karp

University of California at Berkeley

Abstract: A large class of computational problems involve the determination of properties of graphs, digraphs, integers, arrays of integers, finite families of finite sets, boolean formulas and elements of other countable domains. Through simple encodings from such domains into the set of words over a finite alphabet these problems can be converted into language recognition problems, and we can inquire into their computational complexity. It is reasonable to consider such a problem satisfactorily solved when an algorithm for its solution is found which terminates within a number of steps bounded by a polynomial in the length of the input. We show that a large number of classic unsolved problems of covering, matching, packing, routing, assignment and sequencing are equivalent, in the sense that either each of them possesses a polynomial-bounded algorithm or none of them does.

1. INTRODUCTION

All the general methods presently known for computing the chromatic number of a graph, deciding whether a graph has a Hamilton circuit, or solving a system of linear inequalities in which the variables are constrained to be 0 or 1, require a combinatorial search for which the worst case time requirement grows exponentially with the length of the input. In this paper we give theorems which strongly suggest, but do not imply, that these problems, as well as many others, will remain intractable perpetually.

†This research was partially supported by National Science Foundation Grant GJ-474.

We are specifically interested in the existence of algorithms that are guaranteed to terminate in a number of steps bounded by a polynomial in the length of the input. We exhibit a class of well-known combinatorial problems, including those mentioned above, which are equivalent, in the sense that a polynomial-bounded algorithm for any one of them would effectively yield a polynomial-bounded algorithm for all. We also show that, if these problems do possess polynomial-bounded algorithms then all the problems in an unexpectedly wide class (roughly speaking, the class of problems solvable by polynomial-depth backtrack search) possess polynomial-bounded algorithms.

The following is a brief summary of the contents of the paper. For the sake of definiteness our technical development is carried out in terms of the recognition of languages by one-tape Turing machines, but any of a wide variety of other abstract models of computation would yield the same theory. Let Σ^* be the set of all finite strings of 0's and 1's. A subset of Σ^* is called a <u>language</u>. Let $\mathcal{P}$ be the class of languages recognizable in polynomial time by one-tape deterministic Turing machines, and let $\mathcal{NP}$ be the class of languages recognizable in polynomial time by one-tape nondeterministic Turing machines. Let Π be the class of functions from Σ^* into Σ^* computable in polynomial time by one-tape Turing machines. Let L and M be languages. We say that $L \propto M$ (L <u>is reducible to</u> M) if there is a function $f \in \Pi$ such that $f(x) \in M \Leftrightarrow x \in L$. If $M \in \mathcal{P}$ and $L \propto M$ then $L \in \mathcal{P}$. We call L and M equivalent if $L \propto M$ and $M \propto L$. Call L <u>(polynomial) complete</u> if $L \in \mathcal{NP}$ and every language in $\mathcal{NP}$ is reducible to L. Either all complete languages are in $\mathcal{P}$, or none of them are. The former alternative holds if and only if $\mathcal{P} = \mathcal{NP}$.

The main contribution of this paper is the demonstration that a large number of classic difficult computational problems, arising in fields such as mathematical programming, graph theory, combinatorics, computational logic and switching theory, are complete (and hence equivalent) when expressed in a natural way as language recognition problems.

This paper was stimulated by the work of Stephen Cook (1971), and rests on an important theorem which appears in his paper. The author also wishes to acknowledge the substantial contributions of Eugene Lawler and Robert Tarjan.

2. THE CLASS $\mathcal{P}$

There is a large class of important computational problems which involve the determination of properties of graphs, digraphs, integers, finite families of finite sets, boolean formulas and elements of other countable domains. It is a reasonable working hypothesis, championed originally by Jack Edmonds (1965) in connection with problems in graph theory and integer programming, and by now widely accepted, that such a problem can be regarded as tractable if and only if there is an algorithm for its solution whose running time is bounded by a polynomial in the size of the input. In this section we introduce and begin to investigate the class of problems solvable in polynomial time.

We begin by giving an extremely general definition of "deterministic algorithm", computing a function from a countable domain D into a countable range R.

For any finite alphabet A, let A^* be the set of finite strings of elements of A; for $x \in A^*$, let $lg(x)$ denote the length of x.

A <u>deterministic algorithm</u> $\mathcal{A}$ is specified by:

a countable set D (the <u>domain</u>)
a countable set R (the <u>range</u>)
a finite alphabet Δ such that $\Delta^* \wedge R = \phi$
an <u>encoding function</u> $E: D \to \Delta^*$
a <u>transition function</u> $\tau: \Delta^* \to \Delta^* \cup R$.

The <u>computation</u> of $\mathcal{A}$ on input $x \in D$ is the unique sequence $y_1, y_2, \ldots$ such that $y_1 = E(x)$, $y_{i+1} = \tau(y_i)$ for all i and, if the sequence is finite and ends with y_k, then $y_k \in R$. Any string occurring as an element of a computation is called an <u>instantaneous description</u>. If the computation of $\mathcal{A}$ on input x is finite and of length $t(x)$, then $t(x)$ is the <u>running time</u> of $\mathcal{A}$ on input x. $\mathcal{A}$ is <u>terminating</u> if all its computations are finite. A terminating algorithm $\mathcal{A}$ computes the function $f_{\mathcal{A}}: D \to R$ such that $f_{\mathcal{A}}(x)$ is the last element of the computation of $\mathcal{A}$ on x.

If $R = \{ACCEPT, REJECT\}$ then $\mathcal{A}$ is called a <u>recognition algorithm</u>. A recognition algorithm in which $D = \Sigma^*$ is called a <u>string recognition algorithm</u>. If $\mathcal{A}$ is a string recognition algorithm then the <u>language recognized by</u> $\mathcal{A}$ is $\{x \in \Sigma^* | f_{\mathcal{A}}(x) = ACCEPT\}$. If $D = R = \Sigma^*$ then $\mathcal{A}$ is called a <u>string mapping algorithm</u>. A terminating algorithm $\mathcal{A}$ with domain $D = \Sigma^*$ <u>operates in polynomial time</u> if there is a polynomial $p(\cdot)$ such that, for every $x \in \Sigma^*$, $t(x) \le p(lg(x))$.

To discuss algorithms in any practical context we must specialize the concept of deterministic algorithm. Various well known classes of string recognition algorithms (Markov algorithms, one-tape Turing machines, multitape and multihead Turing machines, random access machines, etc.) are delineated by restricting the functions E and τ to be of certain very simple types. These definitions are standard [Hopcroft & Ullman (1969)] and will not be repeated here. It is by now commonplace to observe that many such classes are equivalent in their capability to recognize languages; for each such class of algorithms, the class of languages recognized is the class of recursive languages. This invariance under changes in definition is part of the evidence that recursiveness is the correct technical formulation of the concept of decidability.

The class of languages recognizable by string recognition algorithms which operate in polynomial time is also invariant under a wide range of changes in the class of algorithms. For example, any language recognizable in time $p(\cdot)$ by a multihead or multitape Turing machine is recognizable in time $p^2(\cdot)$ by a one-tape Turing machine. Thus the class of languages recognizable in polynomial time by one-tape Turing machines is the same as the class recognizable by the ostensibly more powerful multihead or multitape Turing machines. Similar remarks apply to random access machines.

Definition 1. $\mathcal{P}$ is the class of languages recognizable by one-tape Turing machines which operate in polynomial time.

Definition 2. Π is the class of functions from Σ^* into Σ^* defined by one-tape Turing machines which operate in polynomial time.

The reader will not go wrong by identifying $\mathcal{P}$ with the class of languages recognizable by digital computers (with unbounded backup storage) which operate in polynomial time and Π with the class of string mappings performed in polynomial time by such computers.

Remark. If $f: \Sigma^* \to \Sigma^*$ is in Π then there is a polynomial $p(\cdot)$ such that $\lg(f(x)) \leq p(\lg(x))$.

We next introduce a concept of reducibility which is of central importance in this paper.

Definition 3. Let L and M be languages. Then $L \propto M$ (L is reducible to M) if there is a function $f \in \Pi$ such that $f(x) \in M \Leftrightarrow x \in L$.

Lemma 1. If $L \propto M$ and $M \in \mathcal{P}$ then $L \in \mathcal{P}$.

Proof. The following is a polynomial-time bounded algorithm to decide if $x \in L$: compute $f(x)$; then test in polynomial time whether $f(x) \in M$.

We will be interested in the difficulty of recognizing subsets of countable domains other than Σ^*. Given such a domain D, there is usually a natural one-one encoding $e: D \to \Sigma^*$. For example we can represent a positive integer by the string of 0's and 1's comprising its binary representation, a 1-dimensional integer array as a list of integers, a matrix as a list of 1-dimensional arrays, etc.; and there are standard techniques for encoding lists into strings over a finite alphabet, and strings over an arbitrary finite alphabet as strings of 0's and 1's. Given such an encoding $e: D \to \Sigma^*$, we say that a set $T \subseteq D$ is <u>recognizable in polynomial time</u> if $e(T) \in \mathcal{P}$. Also, given sets $T \subseteq D$ and $U \subseteq D'$, and encoding functions $e: D \to \Sigma^*$ and $e': D' \to \Sigma^*$ we say $T \propto U$ if $e(T) \propto e'(U)$.

As a rule several natural encodings of a given domain are possible. For instance a graph can be represented by its adjacency matrix, by its incidence matrix, or by a list of unordered pairs of nodes, corresponding to the arcs. Given one of these representations, there remain a number of arbitrary decisions as to format and punctuation. Fortunately, it is almost always obvious that any two "reasonable" encodings e_0 and e_1 of a given problem are equivalent; i.e., $e_0(S) \in \mathcal{P} \Leftrightarrow e_1(S) \in \mathcal{P}$. One important exception concerns the representation of positive integers; we stipulate that a positive integer is encoded in a binary, rather than unary, representation. In view of the invariance of recognizability in polynomial time and reducibility under reasonable encodings, we discuss problems in terms of their original domains, without specifying an encoding into Σ^*.

We complete this section by listing a sampling of problems which are solvable in polynomial time. In the next section we examine a number of close relatives of these problems which are not known to be solvable in polynomial time. Appendix 1 establishes our notation.

Each problem is specified by giving (under the heading "INPUT") a generic element of its domain of definition and (under the heading "PROPERTY") the property which causes an input to be accepted.

SATISFIABILITY WITH AT MOST 2 LITERALS PER CLAUSE [Cook (1971)]
INPUT: Clauses $C_1, C_2, \ldots, C_p$, each containing at most 2 literals
PROPERTY: The conjunction of the given clauses is satisfiable; i.e., there is a set $S \subseteq \{x_1, x_2, \ldots, x_n, \bar{x}_1, \bar{x}_2, \ldots, \bar{x}_n\}$ such that

a) S does not contain a complementary pair of literals and
b) $S \cap C_k \neq \phi$, $k = 1, 2, \ldots, p$.

MINIMUM SPANNING TREE [Kruskal (1956)]
INPUT: G, w, W
PROPERTY: There exists a spanning tree of weight $\leq W$.

SHORTEST PATH [Dijkstra (1959)]
INPUT: G, w, W, s, t
PROPERTY: There is a path between s and t of weight $\leq$ W.

MINIMUM CUT [Edmonds & Karp (1972)]
INPUT: G, w, W, s, t
PROPERTY: There is an s,t cut of weight $\leq$ W.

ARC COVER [Edmonds (1965)]
INPUT: G, k
PROPERTY: There is a set $Y \subseteq A$ such that $|Y| \leq k$ and every node is incident with an arc in Y.

ARC DELETION
INPUT: G, k
PROPERTY: There is a set of k arcs whose deletion breaks all cycles.

BIPARTITE MATCHING [Hall (1948)]
INPUT: $S \subseteq Z_p \times Z_p$
PROPERTY: There are p elements of S, no two of which are equal in either component.

SEQUENCING WITH DEADLINES
INPUT: $(T_1,\ldots,T_n) \in Z^n$, $(D_1,\ldots,D_n) \in Z^n$, k
PROPERTY: Starting at time 0, one can execute jobs 1,2,...,n, with execution times T_i and deadlines D_i, in some order such that not more than k jobs miss their deadlines.

SOLVABILITY OF LINEAR EQUATIONS
INPUT: (c_{ij}), (a_i)
PROPERTY: There exists a vector (y_j) such that, for each i, $\sum_j c_{ij}y_j = a_i$.

3. NONDETERMINISTIC ALGORITHMS AND COOK'S THEOREM

In this section we state an important theorem due to Cook (1971) which asserts that any language in a certain wide class NP is reducible to a specific set S, which corresponds to the problem of deciding whether a boolean formula in conjunctive normal form is satisfiable.

Let $P^{(2)}$ denote the class of subsets of $\Sigma^* \times \Sigma^*$ which are recognizable in polynomial time. Given $L^{(2)} \in P^{(2)}$ and a polynomial p, we define a language L as follows:

$$L = \{x \mid \text{there exists } y \text{ such that } \langle x,y\rangle \in L^{(2)} \text{ and } \lg(y) \leq p(\lg(x))\}.$$

We refer to L as the language derived from $L^{(2)}$ by p-bounded existential quantification.

Definition 4. *NP* is the set of languages derived from elements of $P^{(2)}$ by polynomial-bounded existential quantification.

There is an alternative characterization of *NP* in terms of nondeterministic Turing machines. A nondeterministic recognition algorithm *A* is specified by:

a countable set D (the domain)
a finite alphabet Δ such that $\Delta^* \cap \{ACCEPT, REJECT\} = \phi$
an encoding function $E: D \to \Delta^*$
a transition relation $\tau \subseteq \Delta^* \times (\Delta^* \cup \{ACCEPT, REJECT\})$

such that, for every $y_o \in \Delta^*$, the set $\{<y_o,y> \mid <y_o,y> \in \tau\}$ has fewer than k_A elements, where k_A is a constant. A computation of *A* on input $x \in D$ is a sequence $y_1, y_2, \ldots$ such that $y_1 = E(x)$, $<y_i, y_{i+1}> \in \tau$ for all i, and, if the sequence is finite and ends with y_k, then $y_k \in \{ACCEPT, REJECT\}$. A string $y \in \Delta^*$ which occurs in some computation is an instantaneous description. A finite computation ending in ACCEPT is an accepting computation. Input x is accepted if there is an accepting computation for x. If $D = \Sigma^*$ then *A* is a nondeterministic string recognition algorithm and we say that *A* operates in polynomial time if there is a polynomial $p(\cdot)$ such that, whenever *A* accepts x, there is an accepting computation for x of length $\leq p(lg(x))$.

A nondeterministic algorithm can be regarded as a process which, when confronted with a choice between (say) two alternatives, can create two copies of itself, and follow up the consequences of both courses of action. Repeated splitting may lead to an exponentially growing number of copies; the input is accepted if any sequence of choices leads to acceptance.

The nondeterministic 1-tape Turing machines, multitape Turing machines, random-access machines, etc. define classes of nondeterministic string recognition algorithms by restricting the encoding function E and transition relation τ to particularly simple forms. All these classes of algorithms, restricted to operate in polynomial time, define the same class of languages. Moreover, this class is *NP*.

Theorem 1. $L \in NP$ if and only if L is accepted by a nondeterministic Turing machine which operates in polynomial time.

Proof. ⇒ Suppose $L \in NP$. Then, for some $L^{(2)} \in P^{(2)}$ and some polynomial p, L is obtained from $L^{(2)}$ by p-bounded existential quantification. We can construct a nondeterministic machine which first guesses the successive digits of a string y of length $\leq p(\lg(y))$ and then tests whether $\langle x,y\rangle \in L^{(2)}$. Such a machine clearly recognizes L in polynomial time.

$\Leftarrow$ Suppose L is accepted by a nondeterministic Turing machine T which operates in time p. Assume without loss of generality that, for any instantaneous description Z, there are at most two instantaneous descriptions that may follow Z (i.e., at most two primitive transitions are applicable). Then the sequence of choices of instantaneous descriptions made by T in a given computation can be encoded as a string y of 0's and 1's, such that $\lg(y) \leq p(\lg(x))$.

Thus we can construct a deterministic Turing machine T', with $\Sigma^* \times \Sigma^*$ as its domain of inputs, which, on input $\langle x,y\rangle$, simulates the action of T on input x with the sequence of choices y. Clearly T' operates in polynomial time, and L is obtained by polynomial bounded existential quantification from the set of pairs of strings accepted by T'.

The class $\mathcal{NP}$ is very extensive. Loosely, a recognition problem is in $\mathcal{NP}$ if and only if it can be solved by a backtrack search of polynomial bounded depth. A wide range of important computational problems which are not known to be in $\mathcal{P}$ are obviously in $\mathcal{NP}$. For example, consider the problem of determining whether the nodes of a graph G can be colored with k colors so that no two adjacent nodes have the same color. A nondeterministic algorithm can simply guess an assignment of colors to the nodes and then check (in polynomial time) whether all pairs of adjacent nodes have distinct colors.

In view of the wide extent of $\mathcal{NP}$, the following theorem due to Cook is remarkable. We define the satisfiability problem as follows:

SATISFIABILITY
INPUT: Clauses $C_1,C_2,\ldots,C_p$
PROPERTY: The conjunction of the given clauses is satisfiable; i.e., there is a set $S \subseteq \{x_1,x_2,\ldots,x_n;\bar{x}_1,\bar{x}_2,\ldots,\bar{x}_n\}$ such that
a) S does not contain a complementary pair of literals
and b) $S \cap C_k \neq \phi$, $k = 1,2,\ldots,p$.

<u>Theorem 2 (Cook)</u>. If $L \in \mathcal{NP}$ then $L \propto$ SATISFIABILITY.

The theorem stated by Cook (1971) uses a weaker notion of reducibility than the one used here, but Cook's proof supports the present statement.

<u>Corollary 1</u>. $\mathcal{P} = \mathcal{NP} \Leftrightarrow$ SATISFIABILITY $\in \mathcal{P}$.

Proof. If SATISFIABILITY $\in P$ then, for each $L \in NP$, $L \in P$, since $L \propto$ SATISFIABILITY. If SATISFIABILITY $\notin P$, then, since clearly SATISFIABILITY $\in NP$, $P \neq NP$.

Remark. If $P = NP$ then NP is closed under complementation and polynomial-bounded existential quantification. Hence it is also closed under polynomial-bounded universal quantification. It follows that a polynomial-bounded analogue of Kleene's Arithmetic Hierarchy [Rogers (1967)] becomes trivial if $P = NP$.

Theorem 2 shows that, if there were a polynomial-time algorithm to decide membership in SATISFIABILITY then every problem solvable by a polynomial-depth backtrack search would also be solvable by a polynomial-time algorithm. This is strong circumstantial evidence that SATISFIABILITY $\notin P$.

4. COMPLETE PROBLEMS

The main object of this paper is to establish that a large number of important computational problems can play the role of SATISFIABILITY in Cook's theorem. Such problems will be called complete.

Definition 5. The language L is (polynomial) complete if

a) $L \in NP$

and b) SATISFIABILITY $\propto L$.

Theorem 3. Either all complete languages are in P, or none of them are. The former alternative holds if and only if $P = NP$.

We can extend the concept of completeness to problems defined over countable domains other than Σ^*.

Definition 6. Let D be a countable domain, e a "standard" one-one encoding $e: D \to \Sigma^*$ and T a subset of D. Then T is complete if and only if e(D) is complete.

Lemma 2. Let D and D' be countable domains, with one-one encoding functions e and e'. Let $T \subseteq D$ and $T' \subseteq D'$. Then $T \propto T'$ if there is a function $F: D \to D'$ such that

a) $F(x) \in T' \Leftrightarrow x \in T$

and b) there is a function $f \in \Pi$ such that $f(x) = e'(F(e^{-1}(x)))$ whenever $e'(F(e^{-1}(x)))$ is defined.

The rest of the paper is mainly devoted to the proof of the following theorem.

Main Theorem. All the problems on the following list are complete.

1. SATISFIABILITY
COMMENT: By duality, this problem is equivalent to determining whether a disjunctive normal form expression is a tautology.

2. 0-1 INTEGER PROGRAMMING
INPUT: integer matrix C and integer vector d
PROPERTY: There exists a 0-1 vector x such that $Cx = d$.

3. CLIQUE
INPUT: graph G, positive integer k
PROPERTY: G has a set of k mutually adjacent nodes.

4. SET PACKING
INPUT: Family of sets $\{S_j\}$, positive integer ℓ
PROPERTY: $\{S_j\}$ contains ℓ mutually disjoint sets.

5. NODE COVER
INPUT: graph G', positive integer ℓ
PROPERTY: There is a set $R \subseteq N'$ such that $|R| \leq \ell$ and every arc is incident with some node in R.

6. SET COVERING
INPUT: finite family of finite sets $\{S_j\}$, positive integer k
PROPERTY: There is a subfamily $\{T_h\} \subseteq \{S_j\}$ containing $\leq k$ sets such that $\cup T_h = \cup S_j$.

7. FEEDBACK NODE SET
INPUT: digraph H, positive integer k
PROPERTY: There is a set $R \subseteq V$ such that every (directed) cycle of H contains a node in R.

8. FEEDBACK ARC SET
INPUT: digraph H, positive integer k
PROPERTY: There is a set $S \subseteq E$ such that every (directed) cycle of H contains an arc in S.

9. DIRECTED HAMILTON CIRCUIT
INPUT: digraph H
PROPERTY: H has a directed cycle which includes each node exactly once.

10. UNDIRECTED HAMILTON CIRCUIT
INPUT: graph G
PROPERTY: G has a cycle which includes each node exactly once.

11. SATISFIABILITY WITH AT MOST 3 LITERALS PER CLAUSE
INPUT: Clauses $D_1, D_2, \ldots, D_r$, each consisting of at most 3 literals from the set $\{u_1, u_2, \ldots, u_m\} \cup \{\bar{u}_1, \bar{u}_2, \ldots, \bar{u}_m\}$
PROPERTY: The set $\{D_1, D_2, \ldots, D_r\}$ is satisfiable.

12. CHROMATIC NUMBER
INPUT: graph G, positive integer k
PROPERTY: There is a function $\phi: N \to Z_k$ such that, if u and v are adjacent, then $\phi(u) \neq \phi(v)$.

13. CLIQUE COVER
INPUT: graph G', positive integer ℓ
PROPERTY: N' is the union of ℓ or fewer cliques.

14. EXACT COVER
INPUT: family $\{S_j\}$ of subsets of a set $\{u_i, i = 1,2,\ldots,t\}$
PROPERTY: There is a subfamily $\{T_h\} \subseteq \{S_j\}$ such that the sets T_h are disjoint and $\cup T_h = \cup S_j = \{u_i, i = 1,2,\ldots,t\}$.

15. HITTING SET
INPUT: family $\{U_i\}$ of subsets of $\{s_j, j = 1,2,\ldots,r\}$
PROPERTY: There is a set W such that, for each i, $|W \cap U_i| = 1$.

16. STEINER TREE
INPUT: graph G, $R \subseteq N$, weighting function $w: A \to Z$, positive integer k
PROPERTY: G has a subtree of weight $\leq k$ containing the set of nodes in R.

17. 3-DIMENSIONAL MATCHING
INPUT: set $U \subseteq T \times T \times T$, where T is a finite set
PROPERTY: There is a set $W \subseteq U$ such that $|W| = |T|$ and no two elements of W agree in any coordinate.

18. KNAPSACK
INPUT: $(a_1, a_2, \ldots, a_r, b) \in Z^{n+1}$
PROPERTY: $\Sigma a_j x_j = b$ has a 0-1 solution.

19. JOB SEQUENCING
INPUT: "execution time vector" $(T_1, \ldots, T_p) \in Z^p$,
"deadline vector" $(D_1, \ldots, D_p) \in Z^p$
"penalty vector" $(P_1, \ldots, P_p) \in Z^p$
positive integer k
PROPERTY: There is a permutation π of $\{1,2,\ldots,p\}$ such that

$$\left(\sum_{j=1}^{p} [\text{if } T_{\pi(1)} + \cdots + T_{\pi(j)} > D_{\pi(j)} \text{ then } P_{\pi(j)} \text{ else } 0]\right) \leq k .$$

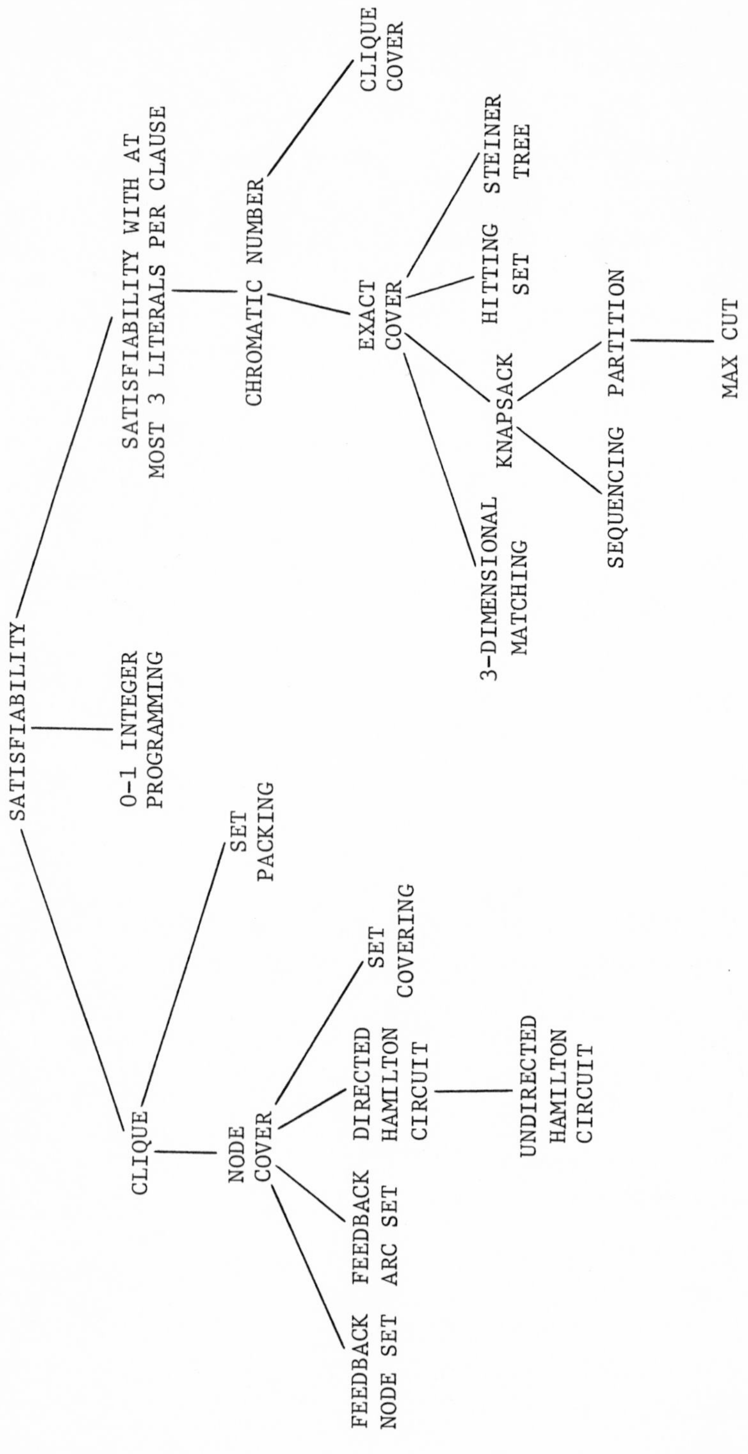

FIGURE 1 - Complete Problems

20. PARTITION
INPUT: $(c_1,c_2,\ldots,c_s) \in Z^s$
PROPERTY: There is a set $I \subseteq \{1,2,\ldots,s\}$ such that $\sum_{h\in I} c_h = \sum_{h\notin I} c_h$.

21. MAX CUT
INPUT: graph G, weighting function $w: A \to Z$, positive integer W
PROPERTY: There is a set $S \subseteq N$ such that

$$\sum_{\substack{\{u,v\}\in A \\ u\in S \\ v\notin S}} w(\{u,v\}) \geq W \quad .$$

It is clear that these problems (or, more precisely, their encodings into Σ^*), are all in *NP*. We proceed to give a series of explicit reductions, showing that SATISFIABILITY is reducible to each of the problems listed. Figure 1 shows the structure of the set of reductions. Each line in the figure indicates a reduction of the upper problem to the lower one.

To exhibit a reduction of a set $T \subseteq D$ to a set $T' \subseteq D'$, we specify a function $F: D \to D'$ which satisfies the conditions of Lemma 2. In each case, the reader should have little difficulty in verifying that F does satisfy these conditions.

SATISFIABILITY $\propto$ 0-1 INTEGER PROGRAMMING

$$c_{ij} = \begin{cases} 1 & \text{if } x_j \in C_i \\ -1 & \text{if } \bar{x}_j \in C_i \\ 0 & \text{otherwise} \end{cases} \qquad \begin{aligned} i &= 1,2,\ldots,p \\ j &= 1,2,\ldots,n \end{aligned}$$

$b_i = 1 -$ (the number of complemented variables in C_i) ,
$i = 1,2,\ldots,p$.

SATISFIABILITY $\propto$ CLIQUE

$N = \{\langle\sigma,i\rangle \mid \sigma \text{ is a literal and occurs in } C_i\}$

$A = \{\{\langle\sigma,i\rangle,\langle\delta,j\rangle\} \mid i \neq j \text{ and } \sigma \neq \bar{\delta}\}$

$k = p$, the number of clauses.

CLIQUE $\propto$ SET PACKING

Assume $N = \{1,2,\ldots,n\}$. The elements of the sets $S_1,S_2,\ldots,S_n$ are those two-element sets of nodes $\{i,j\}$ not in A.

$S_i = \{\{i,j\} \mid \{i,j\} \notin A\}, \quad i = 1,2,\ldots,n$

$\ell = k$.

CLIQUE $\propto$ NODE COVER

G' is the complement of G.
$\ell = |N| - k$

NODE COVER $\propto$ SET COVERING

Assume $N' = \{1,2,\ldots,n\}$. The elements are the arcs of G'. S_j is the set of arcs incident with node j. $k = \ell$.

NODE COVER $\propto$ FEEDBACK NODE SET

$$V = N'$$
$$E = \{\langle u,v\rangle \mid \{u,v\} \in A'\}$$
$$k = \ell$$

NODE COVER $\propto$ FEEDBACK ARC SET

$$V = N' \times \{0,1\}$$
$$E = \{\langle\langle u,0\rangle,\langle u,1\rangle\rangle \mid u \in N'\} \cup \{\langle\langle u,1\rangle,\langle v,0\rangle\rangle \mid \{u,v\} \in A'\}$$
$$k = \ell.$$

NODE COVER $\propto$ DIRECTED HAMILTON CIRCUIT

Without loss of generality assume $A' = Z_m$.

$$V = \{a_1,a_2,\ldots,a_\ell\} \cup \{\langle u,i,\alpha\rangle \mid u \in N' \text{ is incident with } i \in A' \text{ and } \alpha \in \{0,1\}\}$$

$$\begin{aligned}
E = {} & \{\langle\langle u,i,0\rangle,\langle u,i,1\rangle\rangle \mid \langle u,i,0\rangle \in V\} \\
& \cup \{\langle\langle u,i,\alpha\rangle,\langle v,i,\alpha\rangle\rangle \mid i \in A',\ u \text{ and } v \text{ are incident with } i,\ \alpha \in \{0,1\}\} \\
& \cup \{\langle\langle u,i,1\rangle,\langle u,j,0\rangle\rangle \mid u \text{ is incident with } i \text{ and } j \text{ and } \nexists h,\ i < h < j, \text{ such that } u \text{ is incident with } h\} \\
& \cup \{\langle\langle u,i,1\rangle,a_f\rangle \mid 1 \le f \le \ell \text{ and } \nexists h > i \text{ such that } u \text{ is incident with } h\} \\
& \cup \{\langle a_f,\langle u,i,0\rangle\rangle \mid 1 \le f \le \ell \text{ and } \nexists h < i \text{ such that } u \text{ is incident with } h\}\ .
\end{aligned}$$

DIRECTED HAMILTON CIRCUIT $\propto$ UNDIRECTED HAMILTON CIRCUIT

$$N = V \times \{0,1,2\}$$
$$\begin{aligned}
A = {} & \{\{\langle u,0\rangle,\langle u,1\rangle\},\{\langle u,1\rangle,\langle u,2\rangle\} \mid u \in V\} \\
& \cup \{\{\langle u,2\rangle,\langle v,0\rangle\} \mid \langle u,v\rangle \in E\}
\end{aligned}$$

SATISFIABILITY $\propto$ SATISFIABILITY WITH AT MOST 3 LITERALS PER CLAUSE

Replace a clause $\sigma_1 \cup \sigma_2 \cup \cdots \cup \sigma_m$, where the σ_i are literals and $m > 3$, by

$$(\sigma_1 \cup \sigma_2 \cup u_1)(\sigma_3 \cup \cdots \cup \sigma_m \cup \bar{u}_1)(\bar{\sigma}_3 \cup u_1)\cdots(\bar{\sigma}_m \cup u_1) \quad ,$$

where u_1 is a new variable. Repeat this transformation until no clause has more than three literals.

SATISFIABILITY WITH AT MOST 3 LITERALS PER CLAUSE
$\propto$ CHROMATIC NUMBER

Assume without loss of generality that $m \geq 4$.

$$N = \{u_1,u_2,\ldots,u_m\} \cup \{\bar{u}_1,\bar{u}_2,\ldots,\bar{u}_m\} \cup \{v_1,v_2,\ldots,v_m\} \cup \{D_1,D_2,\ldots,D_r\}$$

$$A = \{\{u_i,\bar{u}_i\} \mid i=1,2,\ldots,n\} \cup \{\{v_i,v_j\} \mid i \neq j\} \cup \{\{v_i,x_j\} \mid i \neq j\} \cup \{\{v_i,\bar{x}_j\} \mid i \neq j\} \cup \{\{u_i,D_f\} \mid u_i \notin D_f\} \cup \{\{\bar{u}_i,D_f\} \mid \bar{u}_i \in D_f\}$$

$$k = r+1$$

CHROMATIC NUMBER $\propto$ CLIQUE COVER

G' is the complement of G
$\ell = k$.

CHROMATIC NUMBER $\propto$ EXACT COVER

The set of elements is

$$N \cup A \cup \{\langle u,e,f \rangle \mid u \text{ is incident with } e \text{ and } 1 \leq f \leq k\} \quad .$$

The sets S_j are the following:

for each f, $1 \leq f \leq k$, and each $u \in N$,
$\{u\} \cup \{\langle u,e,f \rangle \mid e \text{ is incident with } u\}$;

for each $e \in A$ and each pair f_1, f_2 such that
$1 \leq f_1 \leq k$, $1 \leq f_2 \leq k$ and $f_1 \neq f_2$
$\{e\} \cup \{\langle u,e,f \rangle, f \neq f_1\} \cup \{\langle v,e,g \rangle \mid g \neq f_2\}$,

where u and v are the two nodes incident with e.

EXACT COVER $\propto$ HITTING SET

The hitting set problem has sets U_i and elements s_j, such that $s_j \in U_i \Leftrightarrow u_i \in S_j$.

EXACT COVER $\propto$ STEINER TREE

$$N = \{n_o\} \cup \{S_j\} \cup \{u_i\}$$
$$R = \{n_o\} \cup \{u_i\}$$
$$A = \{\{n_o,S_j\}\} \cup \{\{S_j,u_i\} \mid u_i \in S_j\}$$
$$w(\{n_o,S_j\}) = |S_j|$$
$$w(\{S_j,u_i\}) = 0$$
$$k = |\{u_i\}| \quad .$$

EXACT COVER $\propto$ 3-DIMENSIONAL MATCHING

Without loss of generality assume $|S_j| \geq 2$ for each j. Let $T = \{\langle i,j \rangle \mid u_i \in S_j\}$. Let α be an arbitrary one-one function from $\{u_i\}$ into T. Let $\pi: T \to T$ be a permutation such that, for each fixed j, $\{\langle i,j\rangle \mid u_i \in S_j\}$ is a cycle of π.

$$U = \{\langle\alpha(u_i),\langle i,j\rangle,\langle i,j\rangle\rangle \mid \langle i,j\rangle \in T\} \cup \{\langle\beta,\sigma,\pi(\sigma)\rangle \mid \text{for all } i,\ \beta \neq \alpha(u_i)\} \quad .$$

EXACT COVER $\propto$ KNAPSACK

Let $d = |\{S_j\}| + 1$. Let $\epsilon_{ji} = \begin{cases} 1 & \text{if } u_i \in S_j \\ 0 & \text{if } u_i \notin S_j \end{cases}$. Let

$r = |\{S_j\}|$, $a_j = \sum \epsilon_{ji} d^{i-1}$ and $b = \dfrac{d^t - 1}{d - 1}$.

KNAPSACK $\propto$ SEQUENCING

$p = r$, $T_i = P_i = a_i$, $D_i = b$.

KNAPSACK $\propto$ PARTITION

$$\begin{aligned} s &= r + 2 \\ c_i &= a_i \ , \quad i = 1,2,\dots,r \\ c_{r+1} &= b + 1 \\ c_{r+2} &= \left(\sum_{i=1}^{r} a_i\right) + 1 - b \end{aligned}$$

PARTITION $\propto$ MAX CUT

$$\begin{aligned} N &= \{1,2,\dots,s\} \\ A &= \{\{i,j\} \mid i \in N,\ j \in N,\ i \neq j\} \\ w(\{i,j\}) &= c_i \cdot c_j \\ W &= \left\lceil \tfrac{1}{4} \sum c_i^2 \right\rceil \end{aligned}$$

Some of the reductions exhibited here did not originate with the present writer. Cook (1971) showed that SATISFIABILITY $\propto$ SATISFIABILITY WITH AT MOST 3 LITERALS PER CLAUSE. The reduction

SATISFIABILITY $\propto$ CLIQUE

is implicit in Cook (1970), and was also known to Raymond Reiter. The reduction

NODE COVER $\propto$ FEEDBACK NODE SET

was found by the Algorithms Seminar at the Cornell University Computer Science Department. The reduction

NODE COVER $\propto$ FEEDBACK ARC SET

was found by Lawler and the writer, and Lawler discovered the reduction

EXACT COVER $\propto$ 3-DIMENSIONAL MATCHING

The writer discovered that the exact cover problem was reducible to the directed traveling-salesman problem on a digraph in which the arcs have weight zero or one. Using refinements of the technique used in this construction, Tarjan showed that

$$\text{EXACT COVER} \propto \text{DIRECTED HAMILTON CIRCUIT}$$

and, independently, Lawler showed that

$$\text{NODE COVER} \propto \text{DIRECTED HAMILTON CIRCUIT} \quad .$$

The reduction

$$\text{DIRECTED HAMILTON CIRCUIT} \propto \text{UNDIRECTED HAMILTON CIRCUIT}$$

was pointed out by Tarjan.

Below we list three problems in automata theory and language theory to which every complete problem is reducible. These problems are not known to be complete, since their membership in NP is presently in doubt. The reader unacquainted with automata and language theory can find the necessary definitions in Hopcroft and Ullman (1969).

EQUIVALENCE OF REGULAR EXPRESSIONS
INPUT: A pair of regular expressions over the alphabet $\{0,1\}$
PROPERTY: The two expressions define the same language.

EQUIVALENCE OF NONDETERMINISTIC FINITE AUTOMATA
INPUT: A pair of nondeterministic finite automata with input alphabet $\{0,1\}$
PROPERTY: The two automata define the same language.

CONTEXT-SENSITIVE RECOGNITION
INPUT: A context-sensitive grammar Γ and a string x
PROPERTY: x is in the language generated by Γ.

First we show that

$$\text{SATISFIABILITY WITH AT MOST 3 LITERALS PER CLAUSE} \propto \text{EQUIVALENCE OF REGULAR EXPRESSIONS} \quad .$$

The reduction is made in two stages. In the first stage we construct a pair of regular expressions over an alphabet $\Delta = \{u_1,u_2,\ldots,u_n,\bar{u}_1,\bar{u}_2,\ldots,\bar{u}_n\}$. We then convert these regular expressions to regular expressions over $\{0,1\}$.

The first regular expression is $\Delta^n\Delta^*$ (more exactly, Δ is written out as $(u_1+u_2+\cdots+u_n+\bar{u}_1+\cdots+\bar{u}_n)$, and Δ^n represents n copies of the expression for Δ concatenated together). The second regular expression is

$$\Delta^n\Delta^* \cup \bigcup_{i=1}^{n} (\Delta^*u_i\Delta^*\bar{u}_i\Delta^* \cup \Delta^*\bar{u}_i\Delta^*u_i\Delta^*) \cup \bigcup_{h=1}^{r} \theta(D_h)$$

where

$$\theta(D_h) = \begin{cases} \Delta^*\bar{\sigma}_1\Delta^* & \text{if } D_h = \sigma_1 \\ \Delta^*\bar{\sigma}_1\Delta^*\bar{\sigma}_2\Delta^* \cup \Delta^*\bar{\sigma}_2\Delta^*\bar{\sigma}_1\Delta^* & \text{if } D_h = \sigma_1 \cup \sigma_2 \\ \Delta^*\bar{\sigma}_1\Delta^*\bar{\sigma}_2\Delta^*\bar{\sigma}_3\Delta^* \cup \Delta^*\bar{\sigma}_1\Delta^*\bar{\sigma}_3\Delta^*\bar{\sigma}_2\Delta^* & \\ \quad \cup \Delta^*\bar{\sigma}_2\Delta^*\bar{\sigma}_1\Delta^*\bar{\sigma}_3\Delta^* \cup \Delta^*\bar{\sigma}_2\Delta^*\bar{\sigma}_3\Delta^*\bar{\sigma}_1\Delta^* & \\ \quad \cup \Delta^*\bar{\sigma}_3\Delta^*\bar{\sigma}_1\Delta^*\bar{\sigma}_2\Delta^* \cup \Delta^*\bar{\sigma}_3\Delta^*\bar{\sigma}_2\Delta^*\bar{\sigma}_1\Delta^* & \\ & \text{if } D_h = \sigma_1 \cup \sigma_2 \cup \sigma_3 \ . \end{cases}$$

Now let m be the least positive integer $\geq \log_2 |\Delta|$, and let ϕ be a 1-1 function from Δ into $\{0,1\}^m$. Replace each regular expression by a regular expression over $\{0,1\}$, by making the substitution $a \rightarrow \phi(a)$ for each occurrence of each element of Δ.

EQUIVALENCE OF REGULAR EXPRESSIONS $\propto$ EQUIVALENCE OF NONDETERMINISTIC FINITE AUTOMATA

There are standard polynomial-time algorithms [Salomaa (1969)] to convert a regular expression to an equivalent nondeterministic automaton. Finally, we show that, for any $L \in NP$,

$$L \underline{\propto} \text{ CONTEXT-SENSITIVE RECOGNITION} \quad .$$

Suppose L is recognized in time $p(\ )$ by a nondeterministic Turing machine. Then the following language $\tilde{L}$ over the alphabet $\{0,1,\#\}$ is accepted by a nondeterministic linear bounded automaton which simulates the Turing machine:

$$\tilde{L} = \{\#^{p(lg(x))}x\#^{p(lg(x))} \mid x \in L\} \quad .$$

Hence $\tilde{L}$ is context-sensitive and has a context-sensitive grammar $\tilde{\Gamma}$. Thus $x \in L$ iff

$$\tilde{\Gamma}, \#^{p(lg(x))}x\#^{p(lg(x))}$$

is an acceptable input to CONTEXT-SENSITIVE RECOGNITION.

We conclude by listing the following important problems in NP which are not known to be complete.

GRAPH ISOMORPHISM
INPUT: graphs G and G'
PROPERTY: G is isomorphic to G'.

NONPRIMES
INPUT: positive integer k
PROPERTY: k is composite.

LINEAR INEQUALITIES
INPUT: integer matrix C, integer vector d
PROPERTY: $Cx \geq d$ has a rational solution.

APPENDIX I

Notation and Terminology Used in Problem Specification

PROPOSITIONAL CALCULUS

$x_1,x_2,\ldots,x_n$	$u_1,u_2,\ldots,u_m$	propositional variables
$\bar{x}_1,\bar{x}_2,\ldots,\bar{x}_n$	$\bar{u}_1,\bar{u}_2,\ldots,\bar{u}_m$	complements of propositional variables
σ,σ_i		literals
$C_1,C_2,\ldots,C_p$	$D_1,D_2,\ldots,D_r$	clauses

$C_k \subseteq \{x_1,x_2,\ldots,x_n,\bar{x}_1,\bar{x}_2,\ldots,\bar{x}_n\}$

$D_\ell \subseteq \{u_1,u_2,\ldots,u_m,\bar{u}_1,\bar{u}_2,\ldots,\bar{u}_m\}$

A clause contains no complementary pair of literals.

SCALARS, VECTORS, MATRICES

Z	the positive integers
Z^p	the set of p-tuples of positive integers
Z_p	the set $\{0,1,\ldots,p-1\}$
k,W	elements of Z
$\langle x,y\rangle$	the ordered pair $\langle x,y\rangle$
(a_i) (y_j) d	vectors with nonnegative integer components
(c_{ij}) C	matrices with integer components

GRAPHS AND DIGRAPHS

$G = (N,A)$ $G' = (N',A')$ finite graphs

N,N' sets of nodes A,A' sets of arcs

s,t,u,v nodes $e,\{u,v\}$ arcs

$(X,\bar{X}) = \{\{u,v\} \mid u \in X \text{ and } v \in \bar{X}\}$ cut

If $s \in X$ and $t \in \bar{X}$, $(X,\bar{X})$ is a s-t cut.

$w: A \to Z$ $w': A' \to Z$ weight functions

The weight of a subgraph is the sum of the weights of its arcs.

$H = (V,E)$ digraph V set of nodes, E set of arcs

$e,\langle u,v\rangle$ arcs

SETS

ϕ	the empty set
$\lvert S\rvert$	the number of elements in the finite set S
$\{S_j\}$ $\{T_h\}$ $\{U_i\}$	finite families of finite sets

PERMUTING INFORMATION IN IDEALIZED TWO-LEVEL STORAGE

Robert W. Floyd

Computer Science Department

Stanford University

Extended Abstract

Assume a computer with a (relatively) slow and large memory consisting of <u>pages</u>, each with a capacity of p <u>records</u>. Available operations for manipulating information in slow memory are limited to selecting two pages, forming a subset, of size at most p, from the union of the two pages, and storing this subset in a third page. Thus the typical operation is

$$a_{i_0} \leftarrow a_{i_1 J_1} \cup a_{i_2 J_2}$$

where the records a_{ij} $(0 \leq j < p)$ are the contents of page a_i, and $a_{iJ} = \{a_{ij} \mid j \in J\}$, with $|J_1| + |J_2| \leq p$. We are assuming that rearrangement of the records within a_{i_0} is carried out at negligible cost during the operation as needed. Where possible, we will limit this rearrangement to circular or linear shifts, so that a_{i_0} is actually formed by masking off subsets of a_{i_1} and a_{i_2}, shifting by designated amounts, and forming the (disjoint) union of the resulting sequences of records.

Given a desired permutation on the information in certain pages of the slow memory, we want to find lower, and constructive upper, bounds on the number of operations to carry out this permutation. We may assume, without changing the required number of operations, that the operations destroy the originals of the records they copy, so that each record only exists in one location. Suppose the permutation Π is specified by two functions Π_1, Π_2 such that

$$a_{ij} = b_{\Pi_1(i,j)\Pi_2(i,j)},$$

where a_{ij} is the current j-th record of the i-th page, and b_{rs} is the s-th record of the r-th page of the desired permutation. In what follows, we shall assume $A = \{a_i\}$, $B = \{b_r\}$.

We define:

$$n_{ir} = |\{j \mid \Pi_1(i,j) = r\}|$$

(the number of records to be sent from the i-th current page to the r-th desired page).

$$e(2^k + \ell) = k \cdot 2^k + (k+2)\ell \quad \text{for} \quad 0 \leq \ell \leq 2^k.$$

(This function is piecewise linear, continuous, and satisfies $e(n) = n \log n$ if $n = 2^k$; e has been chosen to satisfy the recurrence

$$e(1) = 0$$
$$e(x+y) \leq e(x) + e(y) + x + y$$

with equality when $|x-y| \leq 1$.)

$$V = V(A) = \sum_{i,r} e(n_{ir}).$$

One can prove that one operation, transforming the memory A into A', increases V by at most p;

$$V(A') \leq V(A) + p.$$

Thus the required number of operations is at least

$$\frac{V(B) - V(A)}{p}.$$

For the case $p = 2^n$, $\Pi_1(i, j) = j$, $\Pi_2(i, j) = i$, so that $V(A) = 0$, $V(B) = p^2 \log p$, the lower bound $p \log_2 p$, may be achieved (this is the case of transposing a $p \times p$ array, stored one row to a page). More generally, if $p = 2^n$, transposing a $k_1 p \times k_2 p$ array, which may be reduced to transposition of $p \times p$ subarrays, is thus readily shown to be possible in a number of operations equal to the lower bound. To transpose such a $p \times p$ array, where $a_{ij} = b_{ji}$, we interpret the indices as bit strings of length n, representing binary numbers in the range $(0, 2^n-1)$. Define $A^{(0)} = A$. We carry out an iteration n times; after the q-th pass, we have

$$a^{(q)}_{i_1 i_2, j_1 j_2} = a^{(0)}_{i_1 j_2, j_1 i_2},$$

where $i_1 i_2 = i$, $j_1 j_2 = j$, $|i_2| = |j_2| = q$, $i_1 i_2$ denotes concatenation, $|i_2|$ is the length of the string i_2. Here

$$a^{(0)}_{ij} = a_{ij}, \quad a^{(n)}_{ij} = a_{ji} = b_{ij}.$$

During the q-th pass, the page of $A^{(q)}$ whose index is $i_1 \alpha i_2$, with $|\alpha| = 1$ and $|i_2| = q-1$, is formed from the pages of $A^{(q-1)}$ whose indices are $i_1 \alpha i_2$ and $i_1 \bar{\alpha} i_2$; this can be done in a single operation. Thus the total number of operations is n times the number of pages. Furthermore, each operation is readily carried out on most actual machines;

$$a^{(q)}_{i_1 \alpha i_2} = (M_{\alpha q} \cap a^{(q-1)}_{i_1 \alpha i_2} \quad \cup \quad (\bar{M}_{\alpha q} \cap S_{\alpha q}(a^{(q-1)}_{i_1 \bar{\alpha} i_2})),$$

where $M_{\alpha q}$ is an appropriate mask (only $2n$ different masks are needed) and $S_{\alpha q}$ is a shift of 2^q records, right or left according as α is 0 or 1.

Our bounds on matrix transposition for $p = 2^n$ are also achieved by Stone's construction (1971A), which, however, does not lend itself to implementation by shifting and masking on conventional computers.

If p is not a power of two, it can be shown that a $p \times p$ array can be transposed in at most $p \lceil \log_2 p \rceil$ operations. To show this, we first break up the process into four stages, constructing intermediate arrays A', A'', A'''; the relations among them are

$$a_{ij} = a'_{-i,j} = a''_{-i,i+j} = a'''_{j,i+j} = b_{ji}$$

where all the index operations are taken mod p. As A' is simply a renaming of the pages of A, the first step A → A' can be absorbed into the subsequent ones without additional operations. The second step A' → A'' shifts the page a'_k left (circularly) by k places; again, this can be absorbed into subsequent operations. The fourth step A''' → B is like the second, and can be absorbed into the last pass of the third. The third step A'' → A''' shifts the k-th column down (circularly) by k places. This can be done in $\lceil \log_2 p \rceil$ passes, or $p\lceil \log_2 p \rceil$ operations, using the binary expansion of k. On the q-th pass, column k is shifted down by 2^{k-1} places iff 2^{k-1} appears in the binary expansion of q.

$$a^{(q)''}_{k+\ell_2,\ell} = a^{(0)''}_{k\ell}, \quad \text{where } \ell = \ell_1\ell_2, \quad |\ell_2| = q.$$

$$a^{(q+1)''}_{k} = (a^{(q)''}_{k} \cap M_q) \cup (a^{(q)''}_{k-2^q} \cap \bar{M}_q).$$

Because the lower bound for transposition of p × p array is $e(p) \doteq p \log_2 p$, the difference between the lower bound and the constructive upper bound is less than p.

There follows a trace of the transposition of a 4 × 4 array using the first method, and of a 5 × 5 array using the second:

$A = A^{(0)}$	$A^{(1)}$	$A^{(2)} = B = A^T$
a b c d	a e c g	a e i m
e f g h	b f d h	b f j n
i j k ℓ	i m k o	c g k o
m n o p	j n ℓ p	d h ℓ p

A	A'	$A'' = A''^{(0)}$	$A''^{(1)}$	$A''^{(2)}$
a b c d e	a b c d e	a b c d e	a f c h e	a f k p e
f g h i j	u v w x y	v w x y u	v b x d u	v b g ℓ u
k ℓ m n o	p q r s t	r s t p q	r w t y q	r w c h q
p q r s t	k ℓ m n o	n o k ℓ m	n s k p m	n s x d m
u v w x y	f g h i j	j f g h i	j o g ℓ i	j o t y i

$A''^{(3)} = A'''$						$B = A^T$				
a	f	k	p	u		a	f	k	p	u
v	b	g	ℓ	q		b	g	ℓ	q	v
r	w	c	h	m		c	h	m	r	w
n	s	x	d	i		d	i	n	s	x
j	o	t	y	e		e	j	o	t	y

Note that in the second example, an algorithm need not actually construct A', A'', or A'''.

Neither upper nor lower bound is exact in general, although the upper bound is known to be exact when $p = 2^h - 1$. For $p = 5$, the lower bound is 12, the upper is 15, and an optimal algorithm has been shown to require 14 operations.

If the permutation Π is not a transposition, it can still be carried out in $O(w \lceil \log_2 w \rceil)$ operations, where w is the number of pages to be permuted. The method used is to partition the records by successive bits in their desired page index. By information theoretic considerations, most permutations with $w > p$ require $O(w(\log_2 p + \log_2 w))$ operations.

Obviously the above results apply equally, whether (1) the pages are blocks on a disc or drum, the records are in fact records, or (2) the pages are words of internal memory, the records are bits. The latter corresponds to the problem of transposing a Boolean matrix in core memory. The former corresponds to tag sorting of records on a disc memory.

The above results apply to an idealized three-address machine. Work is in progress attempting to apply a similar analysis to idealized single-address machines with fast memories capable of holding two or more pages.

This research was supported in part by the National Science Foundation under grant number GJ-33170X, and the Advanced Research Projects Agency of the Office of the Secretary of Defense under Contract SD-183.

AN N LOG N ALGORITHM TO DISTRIBUTE N RECORDS OPTIMALLY IN A SEQUENTIAL ACCESS FILE

Vaughan R. Pratt

Stanford University

1. GENERALITIES

Imagine an operating system whose daily operation includes accessing various files of a library F, one at a time. Let us assume that the cost of accessing a file is a function only of the location of that file and the location of the file previously accessed. That is, associated with the set L of locations, we have a cost matrix C with c_{ij} representing the cost of accessing the file at location j immediately after visiting location i.

Suppose further that we have collected some location-independent statistics on the usage of the files in the system, say the frequency distribution of the set of possible transitions between files, represented as a matrix T with real entries t_{ij} denoting the proportion of transitions from file (<u>not</u> location) i to file j. Naturally, $\sum_{i \in F} \sum_{j \in F} t_{ij} = 1$.

We wish to minimize file access overhead by suitably assigning files to locations. If a is an assignment function such that file i is assigned to location ai, then the average cost of an access is given by

$$\sum_{i \in F} \sum_{j \in F} c_{ai,aj} t_{ij} \qquad (1)$$

We shall restrict our attention to the special case when there are n files and n locations, one file to a location. In practice, this usually means we are assuming all files to be of the same size, which is a rather serious restriction.

With this restriction, we may rewrite (1) as

$$C \cdot (\pi^T T \pi) \tag{2}$$

where π is any $n \times n$ permutation matrix and $A \cdot B$ denotes $\sum_i \sum_j a_{ij} b_{ij}$. Equivalently, we could write

$$tr(C^T \pi^{-1} T \pi), \tag{3}$$

where tr(A) is Σa_{ii}. Expression (3) was used by Zalesskii (1965A) and (2) by Harper (1970A), both of whom pointed out that if we could find efficiently the π that minimized these expressions, even for the special case when the elements of T and C are non-negative, we could then solve the traveling salesman problem efficiently. But as Karp (1972A) shows elsewhere in these pages, if we could do this, then we could solve in polynomial time a remarkably large class of heretofore difficult problems. Thus we have:

Remark. The general form of the allocation problem, even when restricted to the case of all records being of equal size, is intractable in the sense of Cook (1971A).

2. SPECIAL CASES

Fortunately, there are some positive results for several nontrivial special cases of the problem, mostly due to a theorem of Hardy, Littlewood, and Polyá (1926A). We state and prove their theorem here, partly because we need to explain a useful generalization of the theorem due to Knuth (1972B), and partly because we have a more concise way of analyzing the change in expected cost than in the debugged proof [Hardy, Littlewood, and Polyá (1934A)].

Definition. A vector A of dimension n is in alternating order when $a_1 \le a_n \le a_2 \le a_{n-1} \le a_3 \le a_{n-2} \le \dots$.

Theorem (Hardy-Littlewood-Pólya-Knuth). Let f be a real function, monotonic nondecreasing in the integer domain $[0, \lfloor \frac{n+1}{2} \rfloor]$ and satisfying $f(i) \geq f(n-i)$ for $1 > \lfloor \frac{n+1}{2} \rfloor$. Let P, Q be row vectors of dimension n over the non-negative reals. Let C be an $n \times n$ matrix with entries $c_{ij} = f(|i-j|)$. Then $C \cdot (\pi_1^T P^T Q \pi_2)$, or $C \cdot ((P\pi_1)^T Q\pi_2)$, is minimized over all permutations π_1, π_2 when $P\pi_1$ and $Q\pi_2$ are both in alternating order.

Proof. We exhibit an operation that gradually permutes P and Q into alternating order without increasing $C \cdot ((P\pi_1)^T Q\pi_2)$ at any step.

Let h(R) be the vector formed from the vector R of dimension m by interchanging r_i and r_{m-i+1} when $r_i > r_{m-i+1}$, $i \leq m/2$. Let h'(R) be as for h, except that instead, r_{m-i+2} and r_i are interchanged when $r_{m-i+2} > r_i$, r_1 being ignored. (Below, m will take on only the values n and n+1.)

It is easy to see that for sufficiently large k, $(h'h)^k P$ and $(h'h)^k Q$ are both in alternating order.

It remains to be shown that for any $k \geq 0$, $C \cdot ((h'h)^k P)^T ((h'h)^k Q) \leq C \cdot P^T Q$.

There are two sets of elements of P and Q, one (M) whose members move when h is applied to both P and Q, and one ($\overline{M}$) whose members do not. Within each set, elements retain their relative positions (to within a reflection) during the application of h, since everything in M is just reflected about the center. The expression we are minimizing can be considered as the sum of the interactions $c_{ij} p_i q_j$ between all pairs p_i and q_j. Because $c_{ij} = f(|i-j|)$, the only interactions that h can affect are those between one element from M and one from $\overline{M}$, that is, the mixed interactions. (When m is odd, the two central elements p_c and q_c may be considered as belonging to either M or $\overline{M}$, and thus are not considered to participate in mixed interactions.)

Group the mixed interactions into sets of four, involving p_i, p_{m-i+1}, q_j, and q_{m-j+1}, for $1 \leq i, j \leq \lfloor m/2 \rfloor$. The contribution to the expected access time from each such set is initially

$$p_i q_j f|i-j| - p_i q_{m-j+1} f(m-j+1-i) - p_{m-i+1} q_j f(m-j+1-i) - p_{m-i+1} q_{m-j+1} f|i-j| \quad (4)$$

and after applying h, it becomes

$$p_{m-i+1} q_j f|i-j| - p_{m-i+1} q_{m-j+1} f(m-j+1-i) - p_i q_j f(m-j+1-i) - p_i q_{m-j+1} f|i-j|, \quad (5)$$

regardless of whether the p's or the q's were in M. The net increase of this contribution, due to h, is then (5) minus (4), which reduces to

$$(p_i - p_{m-i+1})(q_j - q_{m-j+1})(f(m-(i+j)+1) - f|i-j|). \quad (6)$$

Exactly one of the first two factors is non-positive, corresponding to the pair of elements from $\overline{M}$. The third factor is always non-negative; to see this, first recall that $1 \leq i, j \leq \lfloor m/2 \rfloor$, whence $0 \leq |i-j| \leq \lfloor m/2 \rfloor - 1$. Also,

$$m - (i+j) + 1 = m + |i-j| - 2.\max(i, j) + 1 > |i-j|$$

(since $i, j \leq \lfloor m/2 \rfloor$).

Hence, when $m - (i+j) + 1 \leq \frac{n+1}{2}$, the monotonicity of f in that range, together with the premise that $m = n$ or $n+1$ only, guarantees that the third term is positive. For $m - (i+j) + 1$ greater than $\frac{n+1}{2}$, we have

$$\begin{aligned} f(m-(i+j)+1) &\geq f(n-m+i+j-1) && \text{(since here } f(i) \geq f(n-i)\text{)} \\ &\geq f(n+1-m+|i-j|) && \text{(monotonicity of } f;\ i, j \geq 1\text{)} \\ &\geq f|i-j| && (m = n \text{ or } n+1 \text{ only}) \end{aligned}$$

and so the third factor is still non-negative.

Since one factor is non-positive and the other two are non-negative, the whole of (6) is non-positive. Hence applying h to P and Q simultaneously cannot increase the expected cost.

To see that the same is true of h', redefine h' equivalently as $h'(P) = (CDR(h(0 \text{ CONS } P^R)))^R$, where $P^R = (p_n, p_{n-1}, \ldots, p_2, p_1)$, $(0 \text{ CONS } P) = (0, p_1, p_2, \ldots, p_n)$, and $CDR(P) = (p_2, p_3, \ldots, p_n)$. It is an easy exercise to show that no operation in this definition of h' can increase the expected access cost. (Note that the 0 will not move when h is applied, and so the CDR merely annihilates it. This, incidentally, is where we need $m = n+1$.) Q. E. D.

It is perhaps worth remarking that when the inequalities in the conditions imposed by the theorem on f are strict, the only optimal arrangements are when P and Q, or P^R and Q^R, are in alternating order.

3. APPLICATIONS

We now apply this theorem to some assignment problems. As one might expect from the conditions of the theorem, we shall assume everywhere that the access probabilities for the files are independent, that is, T may be decomposed into $P^T Q$.

Suppose we have a library of n files with independent probabilities of access $P = (p_1, p_2, \ldots, p_n)$. We want to minimize the expected cost of transitions within the library so we may take Q to be P. Suppose further that the files are to be stored one to a cylinder on a disk, or one to a block on magnetic tape (assuming the rewind capability is not to be used, which in practice would be a reasonable assumption for small libraries involving a hundred feet or so of tape, in which rewinding is slower than backspacing). It is also plausible that (to a good approximation) the time taken to travel between two points is a monotonic increasing function of their separation, but is otherwise independent of the absolute positions of the locations. Taking the cost to be the time then satisfies the conditions imposed on f by the theorem. The above assumptions, together with the main theorem, now guarantee that putting the library in alternating order minimizes the expected cost of an access, since Q is automatically put in alternating order when P is.

The next two problems deal with cyclic locations. First, there are heavy rotating devices such as disks and drums; we wish to optimally arrange the contents of a cylinder. The reader should have no trouble convincing himself that all permutations are equally bad when the cost is a linear function of the angular separation of files. (Simply consider going from p_i to p_j and back again.

Secondly, and more interestingly, there are non-inertial cyclic access devices such as doubly-linked lists in random access storage, or Charge-Coupled Devices (CCD's) and magnetic bubble memories. Here (assuming only one track) the cost of getting from location i to j increases with $|i-j|$, until $|i-j|$ represents half the way round, beyond which point the cost then starts to decrease, because we may then take the obvious short cut. But this cost function also satisfies the conditions of the theorem, and (perhaps more surprisingly than for the linear case) alternating order is still optimal.

So far we have only considered one-dimensional problems. The more usual problem is two-dimensional, e.g., in a disk there is the question of choosing a cylinder as well as a location within that cylinder. Again there is a one-way and a two-way case. At first glimpse, one might suppose that for the one-way case one could combine the results for allocation by cylinders (alternating order) with those for allocating within one-way cylinders, namely by distributing the most probable records arbitrarily around the middle cylinder (that is, midway from center to edge of disk), and then similarly filling the next two cylinders on each side, and so on. The catch is that the cost function is a discontinuous function of the angular separation of files when those files are not in the same cylinder, since if the head has not reached the destination cylinder by the time the destination file is in position, we must wait for another revolution. More precisely, if the head takes time rad to travel to the destination cylinder, and the disk takes time rot to rotate to the destination file, and rev is the time for a complete revolution, then the composite time is

$$\left\lceil \frac{\text{rad-rot}}{\text{rev}} \right\rceil \times \text{rev} + \text{rot}, \qquad (7)$$

assuming rot < rev. For users wanting a stopgap algorithm pending the solution of this problem, I suggest using the first-glimpse algorithm above, except that files should be positioned within cylinders in alternating order, with the maximal element of the middle cylinder diametrically opposite the maximal elements of the remaining cylinders, when the probabilities of the three inmost cylinders clearly dominate the rest, or if all probabilities are relatively high, the maximal elements of the cylinders an even number of cylinders away from the middle cylinder perhaps should be aligned with that of the middle cylinder, leaving the other maximal elements still diametrically opposite.

The general form of the two-way two-dimensional problem seems rather difficult. While it is obvious that the solution will resemble the one-dimensional solution (more probable files grouped together), I have nothing more concrete to suggest than the algorithm of the previous paragraph, except with all maximal elements aligned since there is no discontinuity in this problem. However, a special case of this problem does yield to analysis, namely when the cylinder for each file is pre-ordained, leaving only the order within cylinders to be determined. Each cylinder should be arranged in alternating order. Since the expected cost may be decomposed into interactions between pairs of (not necessarily distinct) cylinders, we may now use the more general form of the theorem, in which $P \neq Q$, to show that these interactions are each minimized independently by using alternating order (as before, the maximal elements should be aligned). Hence, the expected cost is also minimized. While this result may not of itself be very useful for allocation problems, it does tell us that the general problem can be reduced to that of finding the right cylinder for each file.

4. CONCLUSIONS

The situation for allocating equal-sized files with independent probabilities of access to one-dimensional memories is well understood, both for the linear and cyclic cases. The corresponding situation for two-dimensional memories is only partially understood. When the probabilities are not independent, the problem seems very much harder; this problem corresponds to the chassis wiring problem, where the number of wires between plugs corresponds to the probability of a transition between two files plus the probability of the reverse transition (still assuming that the cost function is symmetrical). T. C. Hu conjectures (conversation) that when the cost is a linear function of the separation, the one-dimensional version of this problem can be solved in time $O(n^5)$. When the cost matrix C is arbitrary, we showed that the problem became as intractable as the class of problems discussed by Karp (1972B). Finally, we should remark that Grossman and Silverman (1971A) have independently come up with the restricted version of the Hardy et al theorem that applies to the second paragraph, Section 3.

TOWARD A LOWER BOUND FOR SORTING NETWORKS

David C. Van Voorhis

International Business Machines Corporation

Advanced Systems Development Division Laboratory, Los Gatos, Calif.

INTRODUCTION

A sorting network for N items, or an N-sorter, is a circuit with inputs I = $\{i_1, i_2, \ldots, i_N\}$ and outputs $\emptyset = \{o_1, o_2, \ldots, o_N\}$, such that $\emptyset$ is a monotonically increasing permutation of I. Sorting networks can be constructed using ranks of a basic cell called a comparator, which is essentially a 2-sorter. For example, the 4-sorter depicted in Fig. 1 employs five comparators, which all emit their larger input on their higher output lead.

In this paper we derive a lower bound for S(N), the minimum number of comparators required by an N-sorter.

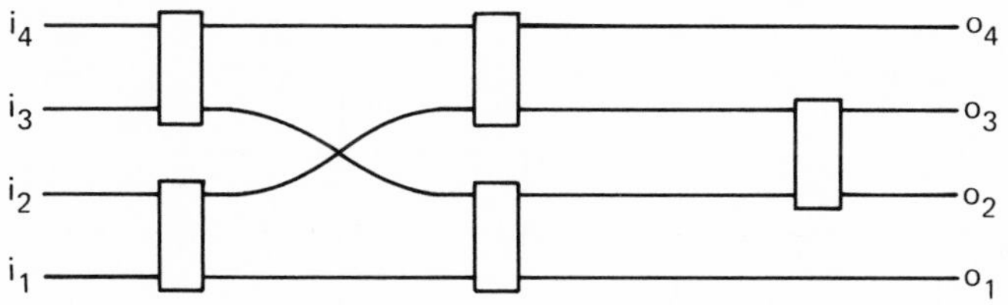

Fig. 1 4-Sorter

PRUNING AN N-SORTER NETWORK

The following procedure, suggested by Green(1970A), can be used to "prune" comparators from an N-sorter to achieve an (N-1)-sorter. If we assign to any input lead i_j a value higher than all other inputs, then this input value follows a unique path p_j from i_j to o_N, becoming the higher output of all comparators traversed. For example, in Fig. 2(a) we show the path from i_3 to o_8 in the 8-sorter designed by Batcher(1968A). Removing the comparators along p_j separates the original N-sorter into two subnetworks: a wire from i_j to o_N and a network with inputs $\{i_1, i_2, \ldots, i_{j-1}, i_{j+1}, \ldots, i_N\}$, and outputs $\{o_1, o_2, \ldots, o_{N-1}\}$, which is an (N-1) sorter, as shown in Fig. 2(b).

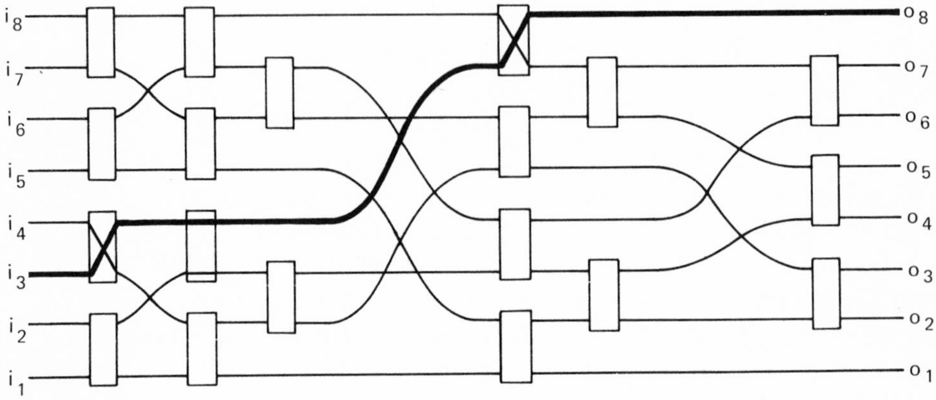

(a) Path from i_3 to o_8

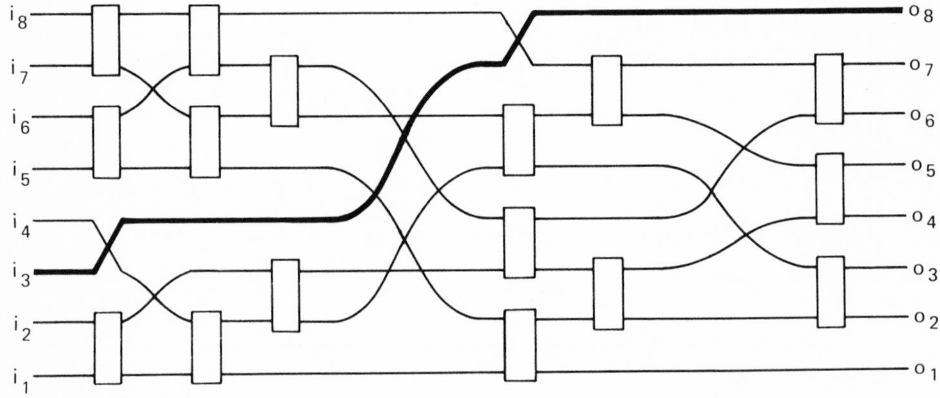

(b) 7-Sorter Network

Fig. 2 Pruning an 8-Sorter Network

Similarly, we can apply to input leads $i_{j1}, i_{j2}, \ldots, i_{jk}$, $k \geq 1$ distinct values higher than the remaining N-k input values, and follow their paths to $o_{N-k+1}, \ldots, o_N$. Pruning the comparators traversed by these paths separates the N-sorter into k wires and an (N-k)-sorter. Let p(k,T) represent the greatest number of comparators that can be pruned from N-sorter T for any of the $\binom{N}{k}$ different choices of k input leads. Then we define

$$P(k,N) = \min_T \; p(k,T), \tag{1}$$

where T ranges over all N-sorters. That is, P(k,N) represents the greatest number of comparators that we can guarantee to eliminate from every N-sorter by selecting $i_{j1}, i_{j2}, \ldots, i_{jk}$ judiciously. It follows immediately that

$$S(N) \geq S(N\text{-}k) + P(k,N). \tag{2}$$

For any N-sorter network T, the paths p_j, $1 \leq j \leq N$, from i_j to o_N together form a subnetwork of T, which we call the MAX subnetwork. The MAX subnetwork of Batcher's 8-sorter is indicated in Fig. 3. It has been observed [Van Voorhis (1972A)] that the MAX subnetwork of any N-sorter is a binary tree with N leaves (the input leads) and N-1 branch nodes (the comparators) rooted at o_N. (That is, o_N is the higher output of the root comparator.) Therefore, the longest path p_j has length $\geq \lceil \log_2 N \rceil$, so that $P(1,N) \geq \lceil \log_2 N \rceil$. And since it is possible to construct a binary tree whose longest path has length $\lceil \log_2 N \rceil$ exactly, $P(1,N) \leq \lceil \log_2 N \rceil$, i. e.

$$P(1, N) = \lceil \log_2 N \rceil. \tag{3}$$

Using (3) in (2) yields the strongest lower bound previously known for S(N), namely

$$S(N) \geq N(\log_2 N) - N + \emptyset(1). \tag{4}$$

In the next section we derive a bound for P(2,N), which in turn provides an even stronger lower bound for S(N).

BOUNDING P(2,N) *

Let T be any N-sorter. The MAX subnetwork of T, MAX(T), includes N-1 comparators, which we label $c_1, c_2, \ldots, c_{N-1}$. Comparator c_j is a branch node of

*In this section we assume several properties of binary trees which are derived, for example, in Knuth(1968A).

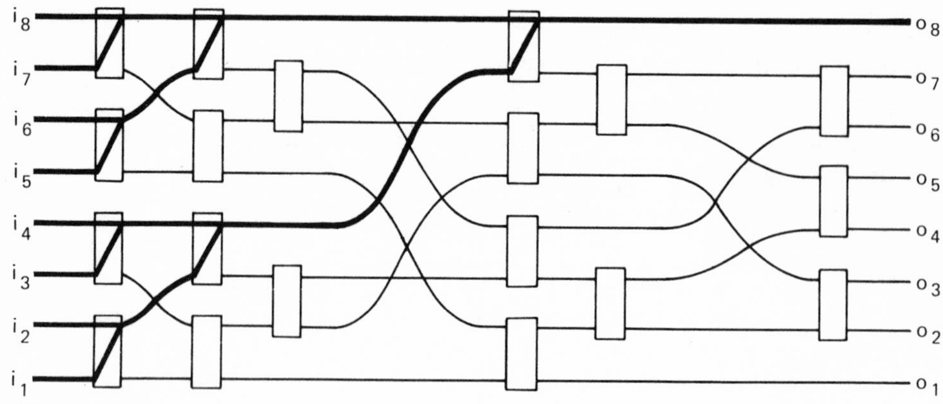

Fig. 3 MAX Subnetwork

the binary tree MAX(T); therefore, it is also the root node of a smaller binary tree C_j. The two subtrees of C_j -- i.e. the portions of MAX(T) leading to the two inputs of c_j -- are denoted $L(C_j)$ and $R(C_j)$.

If we apply the largest input value to a leaf of $L(C_j)$, and we apply the second largest input value to a leaf of $R(C_j)$, these values meet in comparator c_j. The largest input value takes the higher output lead from c_j and proceeds to o_N. Since the second largest input value takes the lower output lead from c_j, T must include a path q_j from the lower output lead of c_j to o_{N-1}. The paths q_j, $1 \leqslant j \leqslant N-1$, together form a binary tree rooted at o_{N-1}, which we call the MAX2 subnetwork. The MAX2 subnetwork for Batcher's 8-sorter is indicated in Fig. 4.

Let p_{j1} be the longest path in MAX(T) that traverses $L(C_j)$, and let p_{j2} be the longest path in MAX(T) that traverses $R(C_j)$. We define $nc(C_j)$ to be the number of comparators from i_{j1} to c_j, plus the number from i_{j2} to c_j, plus the number from c_j to o_N. (Comparator c_j itself is counted exactly once.) The numbers $nc(C_j)$ are included in parentheses following c_j in Fig. 5, for the MAX subnetwork indicated in Fig. 3. If the number of comparators on path q_j is denoted $nc(q_j)$, then the two paths from i_{j1} and i_{j2} through c_j to o_N and o_{N-1} together include $nc(C_j) + nc(q_j)$ comparators. Therefore,

$$p(2, T) = \max_{1 \leqslant j \leqslant N-1} [nc(C_j) + nc(q_j)]. \tag{5}$$

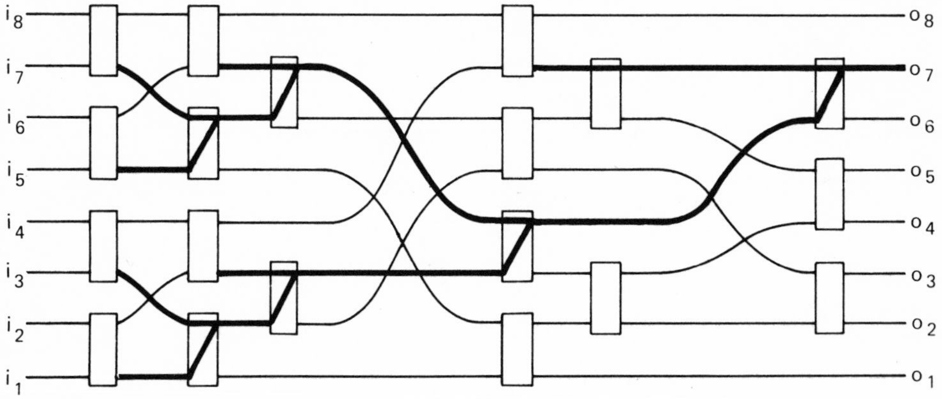

Fig. 4 MAX2 Subnetwork

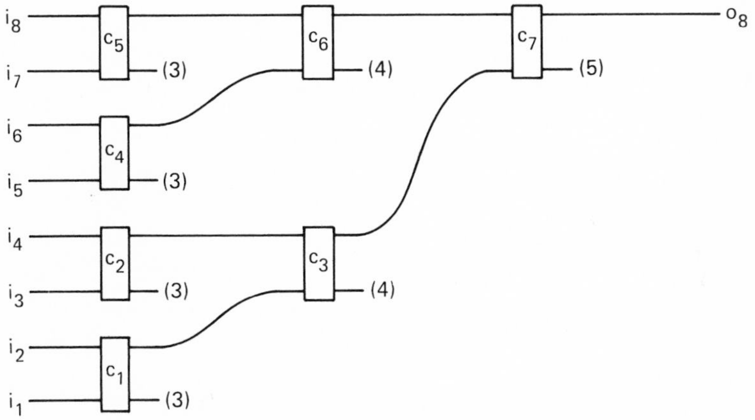

Fig. 5 Symmetric MAX Subnetwork

MAX2(T) is a binary tree, so the path lengths $nc(q_j)$, $1 \leqslant j \leqslant N\text{-}1$, satisfy

$$\sum_{1 \leqslant j \leqslant N-1} 2^{-nc(q_j)} = 1. \tag{6}$$

Equation (5) implies that $nc(q_j) \leqslant p(2,T) - nc(C_j)$, so that

$$\sum_{1 \leqslant j \leqslant N-1} 2^{-[p(2,T)-nc(C_j)]} \leqslant 1. \tag{7}$$

Since $p(2, T)$ is integral, we conclude that

$$p(2,T) \geqslant \lceil \log_2(f(MAX(T))) \rceil, \tag{8}$$

where

$$f(MAX(T)) = \sum_{1 \leqslant j \leqslant N-1} 2^{+nc(C_j)}. \tag{9}$$

Given any N-sorter the bound derived for $p(2, T)$ depends only upon the structure of the subnetwork MAX(T); therefore, we may use (8) and (9) in (1) to obtain

$$P(2,N) \geqslant \min_B \lceil \log_2(f(B)) \rceil, \tag{10}$$

where B ranges over the binary trees with N leaves. Also, defining

$$F(N) = \min_B f(B), \tag{11}$$

we can replace (10) with

$$P(2,N) \geqslant \lceil \log_2(F(N)) \rceil. \tag{12}$$

Knuth(1968A) shows that the number of binary trees with N leaves is $\frac{1}{2N-1}\binom{2N-1}{N}$, so that (11) and (12) provide a useful bound only for small N. However, the following theorem enables us to develop a recursive relation for the function F(N). Let L(B), R(B), and lp(B) represent, respectively, the left and right subtrees of binary tree B, and the length of the longest path of B. If we define $f(\emptyset) = lp(\emptyset) = 0$, where $\emptyset$ is the binary tree with one leaf, then

Theorem 1: $f(B) = 2[f(L(B)) + f(R(B)) + 2^{lp(L(B)) + lp(R(B))}]$. (13)

Proof:

If B has N leaves, then L(B) has k leaves, and R(B) has N-k leaves, where $1 \leqslant k \leqslant N-1$. Let the k-1 comparators in L(B) be labeled $c_{L,1}, c_{L,2}, \ldots, c_{L,k-1}$, and let the N-k-1 comparators in R(B) be labeled $c_{R,1}, c_{R,2}, \ldots, c_{R,N-k-1}$. The N-1 comparators of B can be labeled so that

$$c_j = \begin{cases} c_{L,j}, & 1 \leqslant j \leqslant k-1, \\ c_{R,j-k+1}, & k \leqslant j \leqslant N-2, \end{cases} \tag{14}$$

and c_{N-1} is the comparator joining the roots of L(B) and R(B). With these labels we see that

$$nc(C_j) = nc(C_{L,j}) + 1, \qquad 1 \leqslant j \leqslant k-1, \tag{15}$$

since the path from $c_{L,j}$ to o_N includes one more comparator (c_{N-1}) than the path from $c_{L,j}$ to the root of L(B). Similarly,

$$nc(C_j) = nc(C_{R,j-k+1}) + 1, \qquad k \leqslant j \leqslant N-2. \tag{16}$$

By definition $nc(C_{N-1})$ is the number of comparators on the longest path through $L(C_{N-1}) = L(B)$, plus the number on the longest path through $R(C_{N-1}) = R(B)$, plus the number from c_{N-1} to o_N, i.e.

$$nc(C_{N-1}) = lp(L(B))) + lp(R(B)) + 1. \tag{17}$$

Using (15)-(17) in (9) we obtain

$$f(B) = \sum_{1 \leqslant j \leqslant k-1} 2^{nc(C_{L,j}) + 1} + \sum_{1 \leqslant j \leqslant N-k-1} 2^{nc(C_{R,j}) + 1} + 2^{lp(L(B)) + lp(R(B)) + 1}, \tag{18}$$

which reduces to (13).

Q.E.D.

From the definition $f(\emptyset) = 0$ it follows that $F(1) = 0$. When $L(B)$ has k leaves, $1p(L(B)) \geqslant \lceil \log_2 k \rceil$ and $1p(R(B)) \geqslant \lceil \log_2 (N\text{-}k) \rceil$; therefore, Theorem 1 implies that

Corollary 1: $F(1) = 0$;

$$F(N) \geqslant \min_{1 \leqslant k \leqslant N-1} 2\left[F(k) + F(N\text{-}k) + 2^{\lceil \log_2 k \rceil + \lceil \log_2 (N-k) \rceil}\right]. \quad (19)$$

No closed form solution is known for the recurrence relation given in (19), although when $N = 2^m + q \leqslant 128$ the right-hand-side of (19) achieves its minimum value for $k = \max[2^{m-1}, q]$. (We suggest that this holds in general.) We can discover the asymptotic growth of $F(N)$ by considering the function defined by

$$G(1) = 0;$$

$$G(N) = \min_{1 \leqslant k \leqslant N-1} 2[G(k) + G(N\text{-}k) + k(N\text{-}k)]. \quad (20)$$

$F(N) \geqslant G(N)$, since $2^{\lceil \log_2 x \rceil} \geqslant x$, and

Theorem 2:

The right-hand-side of (20) always achieves a minimum when $k = \lceil \tfrac{1}{2} N \rceil$.

Proof:

We can rewrite (20) as

$$G(1) = 0;$$

$$G(2n+a) = \min_{0 \leqslant r \leqslant n-1} 2[G(n\text{-}r) + G(n+a+r) + (n\text{-}r)(n+a+r)] \quad (21)$$

where $a = 0$ or $a = 1$. Therefore, it suffices to show that

$$G(N) + G(N+a) \leqslant G(N\text{-}r) + G(N+a+r) - r(r+a), \quad 0 \leqslant r \leqslant N\text{-}1. \quad (22)$$

The proof of (22), by induction on N, is straightforward but algebraically tedious, and will be omitted.

Q.E.D.

Immediate consequences of Theorem 2 include

Corollary 2: $G(2^m) = \frac{1}{2}m4^m$. (23)

Corollary 3: $F(2^m) \geqslant \frac{1}{2}m4^m$. (24)

Corollary 4: $P(2,2^m) \geqslant 2m - 1 + \lceil \log_2 m \rceil$. (25)

In the next section we use the bounds derived for P(2,N) to obtain a lower bound for S(N).

BOUNDS FOR S(N)

The strongest lower bound known for S(N) is denoted L(N) and the strongest upper bound known, i.e. the number of comparators actually contained in the most economical N-sorter known, is denoted U(N). In Table 1 we give P(1,N), F(N) P(2,N), L(N), and U(N) for $N \leqslant 16$. For all values of $N \leqslant 16$ except $N = 11$, we have constructed an N-sorter T satisfying $p(2,T) = \lceil \log_2(F(N)) \rceil$. However, we have been able to show that $P(2,11) = 9$ as follows. Our bound for p(2,T) is derived under the assumption that we prune T by removing paths followed by the largest two input values to o_N and o_{N-1}. Clearly it works equally well to prune paths followed by the smallest two input values to o_1 and o_2. It turns out that if T is an 11-sorter such that $f(MAX(T)) = F(N) = 256$, then we can prune two paths to o_1 and o_2 that together include 9 comparators.

The lower bounds L(N), $1 \leqslant N \leqslant 8$, were first proved by Floyd and Knuth (1970A). However, the value $L(7) = 16$ originally required a 20 hour computer check of comparator networks containing 15 comparators, whereas this result follows immediately from $S(5) = 9$ and $P(2,7) = 7$. The upper bounds U(N) are those reported by Green(1970A).

In order to develop a bound for the asymptotic growth of S(N) we define the following two functions L1(N) and L2(N), which are both lower bounds for S(N).

$$\begin{aligned} L1(15) &= L2(15) = 51; & L1(N) &= L1(N-1) + P(1,N), & N &> 16; \\ L1(16) &= L2(16) = 55; & L2(N) &= L2(N-2) + P(2,N), & N &> 16. \end{aligned} \qquad (26)$$

N	P(1,N)	F(N)	P(2,N)	L(N)	U(N)
1	0	0	--------	0	0
2	1	2	1	1	1
3	2	8	3	3	3
4	2	16	4	5	5
5	3	36	6	9	9
6	3	52	6	12	12
7	3	80	7	16	16
8	3	96	7	19	19
9	4	168	8	24	25
10	4	200	8	28	29
11	4	256	9	33	35
12	4	288	9	37	39
13	4	392	9	42	46
14	4	424	9	46	51
15	4	480	9	51	55
16	4	512	9	55	60

Table 1. Bounds for S(N)

When N is a power of 2 we can obtain a convenient formula for L1(N), since

$$L1(2^{m+1}) = L1(2^m) + \sum_{1 \leqslant r \leqslant 2^m} P(1,2^m+r)$$

$$= L1(2^m) + (m+1)2^m, \quad m \geqslant 4. \tag{27}$$

The solution to (27), with boundary condition $L1(16) = 55$, is

$$L1(2^m) = (m-1)2^m + 7, \quad m \geqslant 4. \tag{28}$$

It is impossible to solve for L2(N) since P(2,N) is not known exactly. However, since $F(N) \geqslant G(N)$ and G(N) is strictly increasing, Corollary 2 implies that

$$F(2^m + r) \geqslant \tfrac{1}{2}m4^m + 1, \quad r \geqslant 1. \tag{29}$$

We can use (29) to show that

$$L2(2^{m+1}) \geqslant L2(2^m) + \sum_{1 \leqslant r \leqslant 2^{m-1}} \lceil \log_2(F(2^m+2r)) \rceil$$

$$\geqslant L2(2^m) + \tfrac{1}{2}\lceil \log_2(\tfrac{1}{2}m4^m + 1)\rceil\, 2^m, \quad m \geqslant 4. \tag{30}$$

If we define

$$D(N) = L2(N) - L1(N), \tag{31}$$

then (28) and (30) yield

$$D(2^{m+1}) \geqslant \sum_{4 \leqslant r \leqslant m} (\tfrac{1}{2}\lceil \log_2(\tfrac{1}{2}r4^r+1)\rceil - r-1)2^r, \quad m \geqslant 4. \tag{32}$$

Evidently the right-hand-side of (32) is zero for $m < 8$, and when $m = 2^k \geqslant 8$, (32) can be expressed as

$$D(2^m) \geqslant \tfrac{1}{2}(k-3)2^m - \tfrac{1}{2}(2^8+2^{16}+\ldots+2^{\frac{1}{2}m}), \quad m = 2^k \geqslant 8. \tag{33}$$

Since $L2(2^m)$ is a lower bound for $S(2^m)$, we conclude from (28), (31), and (33) that

$$S(N) \geqslant N(\log_2 N + \tfrac{1}{2}\log_2(\log_2 N)) + \emptyset(N). \tag{34}$$

This represents a significant improvement over the strongest bound previously known for S(N), given by (4), although it still diverges rapidly from the strongest upper bound known [Van Voorhis(1971A)],

$$U(N) = .25N(\log_2 N)^2 - .37N(\log_2 N) + \emptyset(N). \tag{35}$$

CONCLUSION

We have observed that the function P(k,N) can be used to bound S(N) from below. We have shown that $P(1,N) = \lceil \log_2 N \rceil$, and that in most cases $P(2,N) = \lceil \log_2(F(N)) \rceil$, where F(N) satisfies (19). P(1,N) and P(2,N) together provide a lower bound for S(N) which is exact for $N \leqslant 8$, although for large N it diverges rapidly from the strongest upper bound known. The next improvement in the lower bound for S(N) will probably result from an approximation of P(3,N).

ISOMORPHISM OF PLANAR GRAPHS (WORKING PAPER)

J.E. Hopcroft and R.E. Tarjan

Cornell University

Ithaca, New York

ABSTRACT

An algorithm is presented for determining whether or not two planar graphs are isomorphic. The algorithm requires $O(V \log V)$ time, if V is the number of vertices in each graph.

INTRODUCTION

The isomorphism of planar graphs is an important special case of the graph isomorphism problem. It arises in the enumeration of various types of planar graphs and in several engineering disciplines. It is important to consider the planar graph case separately since the more general problem of graph isomorphism is at present intractable. Although good heuristics exist for the general isomorphism problem, all known algorithms have a worst case asymptotic growth rate which is exponential in the number of vertices. In this paper we exhibit an efficient algorithm for testing two planar graphs for isomorphism. The asymptotic running time of the algorithm is bounded by $O(V \log V)$ where V is the number of vertices in the graphs. As a by-product we document a linear tree isomorphism algorithm. Several authors (Edmonds, Scoins, Weinburg and others), have given similar algorithms but no one has published the non-trivial details of implementing the required sorting implied by these algorithms.

Early work on the isomorphism of planar graphs is due primarily to Weinburg who developed efficient algorithms for isomorphism of triply connected planar graphs in time V^2 and for isomorphism of "series parallel" graphs and for trees. His algorithms can clearly be combined to give a polynomial bounded algorithm for the general problem. An improved algorithm for the isomorphism of triply connected planar graphs is given in Hopcroft (1971A) and a V^2 algorithm for isomorphism of planar graphs is given in Hopcroft and Tarjan (1971A).

The paper is divided into five sections. The first section consists of the introduction and certain graph theory terminology. The second section describes an algorithm for partitioning a graph into its unique 3-connected components in time proportional to the number of edges. The third section documents a linear tree isomorphism algorithm. The fourth section describes an algorithm for isomorphism of triply connected planar graphs in time proportional to $V \log V$. The fifth section combines the above algorithms into an isomorphism algorithm for arbitrary planar graphs. The worst case running time of the algorithm grows as $V \log V$.

The remainder of this section is devoted to terminology and notation. We assume that the reader is familiar with the more or less standard definitions of graph theory [Harary (1969A)]. A graph G consists of a finite set of vertices $\mathcal{V}$ and a finite set of edges $\mathcal{E}$. If the edges are unordered pairs of vertices then the graph is *undirected*.[1] If the edges are ordered pairs of vertices, then the graph is *directed*. If (v,w) is a directed edge, then v is called the *tail* and w is called the *head*. A *path*, denoted by $v \overset{*}{\Rightarrow} w$, is a sequence of vertices and edges leading from v to w. A path is *simple* if all its vertices are distinct. A *cycle* is a closed path all of whose edges are distinct and such that only one vertex appears twice.

A *tree* is a connected graph with no cycles. A *rooted tree* is a directed graph satisfying the following three conditions: (1) There is exactly one vertex, called the *root*, which no edge enters. (2) For each vertex in the tree there exists a sequence of directed edges from the root to the vertex. (3) Exactly one edge enters each vertex except the root. A directed edge (v,w) in a tree is denoted $v \rightarrow w$. A path from v to w is denoted by $v \overset{*}{\rightarrow} w$. If $v \rightarrow w$, v is the *father* of w and w is a *son* of v. If $v \overset{*}{\rightarrow} w$, v is an *ancestor* of w and w is a *descendant* of v. Every vertex is an ancestor and a descendant of itself. If v is a vertex in a tree T, then T_v is the subtree of T having as its vertices all the descendants of v in T. Let G be a directed graph. A tree T is a *spanning tree* of G if T is a subgraph of G and T contains all vertices of G.

A graph G is biconnected if for each triple of distinct vertices v,w and a in $\mathcal{V}$ there is a path $p: v \overset{*}{\Rightarrow} w$ such that a is not on the path p. If there is a distinct triple v,w,a such that a is on every path $p:v \overset{*}{\Rightarrow} w$, then a is called an articulation point of G. Let the edges of G be partitioned so that two edges are in the same block of the partition if and only if they belong to a common cycle. Let $G_i = (\mathcal{V}_i, \mathcal{E}_i)$ where $\mathcal{E}_i$ is the set of edges in the ith block of the partition and $\mathcal{V}_i = \{v \mid \exists w \ni (v,w) \in \mathcal{E}_i\}$. Then

(i) Each G_i is biconnected.

(ii) No G_i is a proper subgraph of a biconnected subgraph of G.

(iii) Each vertex of G which is not an articulation point of G occurs exactly once among the $\mathcal{V}_i$ and each articulation point occurs at least twice.

(iv) For each $i,j,i \neq j, \mathcal{V}_i \cap \mathcal{V}_j$ contains at most one vertex; furthermore, this vertex (if any) is an articulation point.

The subgraphs G_i of G are called the biconnected components of G.

A graph is triply connected if for each quadruple of distinct vertices v,w,a,b in $\mathcal{V}$, there is a path $p: v \overset{*}{\Rightarrow} w$ such that neither a nor b is on path p. If there is a quadruple of distinct vertices v,w,a,b in $\mathcal{V}$ such that there is a path $p: v \overset{*}{\Rightarrow} w$ and every such path contains either a or b, then a and b are a biarticulation point pair in G.

An n-gon is a connected graph consisting of a cycle with n edges. An n-bond is a pair of vertices connected by n edges. Strictly speaking an n-bond is not a graph. We introduce it since we intend to find biarticulation point pairs and thereby divide the graph into its triply connected components. When a component is removed, it is replaced by an edge and this process can introduce multiple edges.

By a suitable modification of the above definition (see for instance Tutte (1966A)) or by breaking off only biconnected components, one can insure that the triply connected components are unique.

In deriving time bounds on algorithms we assume a random access model. In order to avoid considering specific details of the model we adopt the following notation. If $\vec{n}$ is a vector and there exist constants k_1,k_2 such that $|t(\vec{n})| \leq k_1|f(\vec{n})| + k_2$, then we write "$t(\vec{n})$ is $O(f(\vec{n}))$".

We make use of several known algorithms. One such algorithm is called a radix sort. We can sort n integers $x_1, x_2, \ldots, x_n$ where each x_i has a value between 1 and n in time $O(n)$ by initializing n buckets. Each x_i is then placed in bucket x_i. Finally the contents of the buckets are removed in order starting with bucket 1.

A graph is stored in the computer using an adjacency structure which consists of a set of adjacency lists, one list for each vertex. The adjacency list for vertex v contains each w such that (v,w) is in $\mathcal{E}$. If G is undirected each edge is represented twice in the adjacency structure. If G is directed, then each edge (v,w) appears only once. The adjacency structure for a graph is not unique and there are as many structures as there are orderings of edges at the vertices.

Representing a graph by its adjacency structure expedites searching the graph. We make use of a particular type of search called a depth-first search. A depth-first search explores a graph by always selecting an edge emanating from the vertex most recently reached which still has unexplored edges.

Let G be an undirected graph. A search of G imposes a direction on each edge of G given by the direction in which the edge is traversed when the search is performed. Thus G is converted into a directed graph G'. The set of edges which lead to a new vertex when traversed during the search defines a spanning tree of G'. In general, the arcs of G' which are not part of the spanning tree interconnect the paths in the tree. However, if the search is depth-first, each edge (v,w) not in the spanning tree connects vertex v with one of its ancestors w. In this case G' is called a palm tree and the arcs of G' not in the spanning tree are called the fronds of G'. An edge (v,w) which is a frond is denoted by $v \dashrightarrow w$. Depth-first search can be implemented in $O(V,E)$ time, using an adjacency structure to give the next edge to be explored from a given vertex.

DETERMINING TRICONNECTIVITY

Let $G = (\mathcal{V},\mathcal{E})$ be a graph with $|\mathcal{V}| = V$ vertices and $|\mathcal{E}| = E$ edges. This section describes an algorithm for determining in $O(V,E)$ time whether G is triconnected. (Older algorithms, such as [Ariyoshi, Shirakana and Hiroshi (1971A)], require $O(V^4)$ time.) The algorithm may be extended to divide a graph into its triconnected components in $O(V,E)$ time, using Tutte's definition [Tutte (1966A)] or some other definition of triconnected components. (Tutte's definition has the advantage that it gives <u>unique</u> components; unique components are necessary to solve the planar isomorphism problem.)

We may assume that $|V| \geq 4$ and that G has no vertices of degree two; if $|V| \leq 4$ or G has a degree two vertex the triconnectivity problem has an immediate answer. Further, we may assume that G is biconnected; [Hopcroft and Tarjan (1971B)] describe a method for dividing a graph into its biconnected components in $O(V,E)$ time.

The triconnectivity algorithm consists of three depth-first searches. The first search constructs a palm tree P for G and calculates information about the fronds of P. An adjacency structure A is constructed for P using the information generated by the first search. The second search uses A to select edges to be explored, and calculates necessary information about P . The third search determines whether a biarticulation point pair exists.

Suppose G is searched in a depth-first fashion and that the vertices of G are numbered in the order they are reached during the search. Let vertices be identified by their numbers. Let P be the palm tree generated by the search. If $v \in \mathcal{V}$, let $LOWPT1(v) = \min(\{v\} \cup \{w \mid v \overset{*}{\rightarrow} \dashrightarrow w\})$. Let

$$LOWPT2(v) = \min[\{v\} \cup (\{w \mid v \overset{*}{\rightarrow} \dashrightarrow w\} - \{LOWPT1(v)\})] .$$

That is, $LOWPT1(v)$ is the lowest vertex reachable from v by traversing zero or more tree arcs in P followed by at most one frond. $LOWPT2(v)$ is the second lowest vertex reachable in this way. We have $LOWPT1(v) < LOWPT2(v)$ unless $LOWPT1(v) = LOWPT2(v) = v$. The numbers and LOWPT values of all vertices may easily be calculated during the first search of G .

If $v \rightarrow w$ in P , let $\phi((v,w)) = LOWPT1(w)$. If $v \dashrightarrow w$ in P, let $\phi((v,w)) = w$. Let A be an adjacency structure for P such that the adjacency lists in A are ordered according to ϕ . (Each entry in an adjacency list corresponds to an edge of P ; the entries must be in order according to the ϕ values of their corresponding edges.) Such an adjacency structure A can be constructed using a single radix sort of the edges of P . See [Tarjan (1972B)]. Furthermore, A depends only on the order of the LOWPT1 values and not on the exact numbering scheme. That is, if the vertices of P are numbered from 1 to V in any manner such that $v \to w$ implies NUMBER(v) < NUMBER(w), and LOWPT1 values are calculated using the new numbers, then the possible adjacency structures A which satisfy the new LOWPT1 ordering are the same as those which satisfy the old ordering. This fact is easy to prove; see [Tarjan (1972B)].

The second search explores the edges in the order given by A, using the same starting vertex as the first search. Vertices are numbered from V to 1 as they are last examined during the search. For this numbering scheme, $v \to w$ implies NUMBER(v) < NUMBER(w); and $v \to w_1$, $v \to w_2$ implies NUMBER(w_1) > NUMBER(w_2), if (v,w_1) is traversed before (v,w_2) during the second search. Henceforth vertices will be identified according to the numbers assigned to them by the second search. LOWPT1 and LOWPT2 values are recalculated using the new numbering. Two other important sets of numbers are needed. If $u \to v$ in P, let $\text{HIGHPT}(v) = \max(\{u\} \cup \{w \mid v \overset{*}{\to} w \;\&\; w \dashrightarrow u\})$. HIGHPT(v) is the highest endpoint of a frond which starts at a descendant of v and ends at the father of v. If v is a vertex, let H(v) be the highest numbered descendant of v. The values of HIGHPT(v) and H(v) for each vertex v are calculated during the second search.

After the second search is completed, we have a palm tree P for G which is ordered according to the adjacency structure A. We also have several sets of numbers associated with the vertices of P. From this information we can determine the biarticulation point pairs of G.

Lemma 1. Let P be a palm tree generated by a depth-first search of a biconnected graph G. Let (a,b) be a biarticulation point pair in G, such that $a < b$. Then $a \overset{*}{\to} b$ in the spanning tree T of P.

Proof. Suppose that b is not a descendant of a in P. If v is a vertex in T, let D(v) be its set of descendants. The subgraph of G with vertices $W = V - D(a) - D(b)$ is connected. If v is any son of a or b, then the vertices in D(v) are adjacent only to vertices in $D(v) \cup W \cup \{a,b\}$. If (a,b) is a biarticulation point pair, either a or b must be an articulation point, which is impossible since G is biconnected.

An elaboration of Lemma 1 gives a necessary and sufficient condition for (a,b) to be a biarticulation point pair. If $v \to w$ and w is the first entry in the adjacency list of v, then w is called the first son of v. If $v \overset{*}{\to} w$ in P, w is a first descendant of v if each vertex except v on the path $v \overset{*}{\to} w$ is a first son of its father. Every vertex is a first descendant of itself.

Lemma 2. Let P be a palm tree generated by a depth-first search of a biconnected graph G. Let LOWPT1 and LOWPT2 be defined as above. Let (a,b) be a biarticulation point pair in G with $a < b$. Then either:

(1) There are distinct vertices $r \neq a,b$ and $s \neq a,b$ such that $b \to r$, $\mathrm{LOWPT1}(r) = a$, $\mathrm{LOWPT2}(r) \geq b$, and s is not a descendant of r. (Pair (a,b) is called a biarticulation point pair of type 1.)

Or:

(2) There is a vertex $r \neq b$ such that $a \to r \overset{*}{\to} b$; b is a first descendant of r; $a \neq 1$; every frond $i \dashrightarrow j$ with $r \leq i < b$ has $a \leq j$; and every frond $i \dashrightarrow j$ with $a < j < b$ and $b \to w \overset{*}{\to} i$ has $\mathrm{LOWPT1}(w) \geq a$. (Pair (a,b) is called a biarticulation point pair of type 2.)

Conversely, any pair of vertices (a,b) which satisfy either (1) or (2) is a biarticulation point pair.

Proof. The converse part of the lemma is easy to prove. To prove the direct part, suppose (a,b) is a biarticulation point pair in G with $a < b$. By Lemma 1, $a \overset{*}{\to} b$ in T, the spanning tree of P. Let $b_1, b_2, \ldots, b_n$ be the sons of b in the order they occur in A_b, the adjacency list of b in A. Let $a \to v \overset{*}{\to} b$. If w is a vertex in P, let $D(w)$ be the set of descendants of w in T. Let $X = D(v) - D(b)$ and $W = V - D(a)$. If $w \neq v$ is a son of a, some vertex in $D(w)$ is adjacent to some vertex in W, since G is biconnected and vertices in $D(w)$ are adjacent only to vertices in $D(w) \cup \{a\} \cup W$. Vertices in $D(b_i)$ are adjacent only to vertices in $D(b_i) \cup \{a,b\} \cup X \cup W$.

If removal of a and b disconnects some $D(b_i)$ from the rest of the graph, then it is easy to show that (a,b) satisfies (1) with $r = b_i$ and s some vertex in the rest of the graph. If this is not the case, then removal of a and b must disconnect X and possibly some of the $D(b_i)$ from W and the rest of the $D(b_i)$. Furthermore, since

$$\mathrm{LOWPT1}(b_1) \leq \mathrm{LOWPT1}(b_2) \leq \cdots \leq \mathrm{LOWPT1}(b_n),$$

there is a $k_0 \geq 1$ such that $W, D(b_1), \ldots, D(b_{k_0})$ are disconnected from $X, D(b_{k_0+1}), \ldots, D(b_n)$. In fact we have

$$k_0 = \max\{i \mid \mathrm{LOWPT1}(b_i) < a\}.$$

Since W and X are not empty, $a \neq 1$ and $v \neq b$. Every frond $i \dashrightarrow j$ with $v \leq i < b$ starts at a vertex in X and hence must satisfy $a \leq j$. Every frond with $a < j < b$ and $b \rightarrow b_k \overset{*}{\rightarrow} i$ must start in some $D(b_k)$ with $k > k_0$ and hence $LOWPT1(b_k) \geq a$. Because G is biconnected, $LOWPT1(v) < a$. If b were not a first descendant of v, some frond in X would lead to a vertex in W. Thus b must be a first descendant of v, and (a,b) is a biarticulation point pair of type 2, with $r = v$ in the definition of a type 2 pair. This gives the direct part of the theorem. Note that a biarticulation point pair may be both of type 1 and of type 2.

Lemma 2 gives an easy criterion for determining the biarticulation point pairs of G. To test for type 1 pairs, we examine each tree arc $b \rightarrow v$ of P and test whether $LOWPT2(v) \geq b$ and either $\neg(LOWPT1(v) \rightarrow b)$ or $LOWPT1(v) \neq 1$ or b has more than one son. If so, (LOWPT1(v),b) is a type 1 pair. Testing for type 2 pairs requires a third search. We keep a stack. Each entry on the stack is a triple (h,a,b) of vertices. The triple denotes that (a,b) is a possible type 2 pair and h is the highest vertex which is connected to vertices in $D(a) - D(b)$ by a path which doesn't pass through a or b. (Vertex $h = H(b_{k_0+1})$ where b_{k_0+1} is defined as in the proof of Lemma 2.)

A depth-first search identical to the second search is performed, and the stack of triples is updated in the following manner: When a frond (v,w) is traversed, all triples (h,a,b) on top of the stack with $w < a$ are deleted. If (h_1,a,b_1) is the last triple deleted, a new triple (h_1,w,b_1) is added to the stack. If no triples are deleted, (v,w,v) is added to the stack.

Whenever we return to a vertex $v \neq 1$ along a tree arc $v \rightarrow w$ during the search, we test the top triple (h,a,b) on the stack to see if $v = a$. If so, (a,b) is a type 2 pair. We also delete all triples (h,a,b) on top of the stack with $HIGHPT(w) > h$. If w is not the first son of v, let H(w) be the highest descendant of w. We delete all triples (h,a,b) on top of the stack with $H(w) \geq b$. Then we delete all triples with $LOWPT1(w) < a$. If no triples are deleted during the latter step and $\neg(LOWPT1(w) \rightarrow v)$, we add the triple (H(w),LOWPT1(w), v) to the stack. Otherwise, if (h,a,b) was the last triple deleted such that $LOWPT1(w) < a$, we add $(\max\{H(w),h\}$, LOWPT1(w), b) to the stack.

If G has one or more type 2 pairs, we will have discovered one of them when the third search is completed.

Lemma 3. The method described above will find a biarticulation point pair if G is not triconnected. On the other hand, if G is triconnected the method described above will not yield a pair of points.

Proof. If G has a pair of type 1 it will be found by the type 1 test; if G has no type 1 pairs the type 1 test will yield no pairs. This follows from (1) in Lemma 2; a vertex w satisfying (1) exists if and only if either $\neg(\mathrm{LOWPT1}(v) \rightarrow b)$ or $\mathrm{LOWPT1}(v) \neq 1$ or b has more than one son.

Suppose now that G has no type 1 pairs. Consider the type 2 test. If (h_1,a_1,b_1) occurs above (h_2,a_2,b_2) in the stack, $a_2 \leq a_1$ and if $a_2 = a_1$ then $b_2 \leq b_1$. Further, if (h,a,b) is deleted from the stack because a frond $v \dashrightarrow w$ is found with $w < a$, then $v < b$. These facts may be proved by induction using the ordering given by A. Every triple (h,a,b) on the stack has $(a \overset{*}{\rightarrow} b)$ and a is a proper ancestor of the vertex currently being examined during the search.

If triple (h,a,b) on the stack is tested and it is found that $v = a \neq 1$ when returning along a tree arc $v \rightarrow w$, it is straightforward to prove by induction that (a,b) is a type 2 pair. Conversely, if (a,b) is a type 2 pair, let $h = H(b_{k_0+1})$, where b_{k_0+1} is defined in the proof of Lemma 2. Let $a \rightarrow v \overset{*}{\rightarrow} b$ and let $i \dashrightarrow j$ be the first frond traversed during the search with $v \leq i \leq h$. Then we may prove by induction that (i,j,i) is placed on the stack, possibly modified, and eventually is selected as a type 2 pair. Thus the tests for type 1 and type 2 pairs correctly determine whether G is triconnected.

Lemma 4. The triconnectivity algorithm requires $O(V,E)$ time.

Proof. The three searches, including the auxiliary calculations, require $O(V,E)$ time. Constructing A requires $O(V,E)$ time if a radix sort is used. Testing for type 1 pairs requires $O(V)$ time. Thus the total time required by the algorithm is linear in V and E.

TREE ISOMORPHISM

Suppose we are given two trees T_1 and T_2, and we wish to discover whether T_1 and T_2 are isomorphic. We may assume that T_1 and T_2 are rooted, since if T_1 and T_2 are not rooted it is possible to select a unique distinguished vertex r_i in each tree T_i and call this vertex the root. This is done by eliminating all vertices of degree one (the leaves) from T_i and repeating this step until either a single vertex (the center of T_i) is left or a single edge (whose endpoints are the bicenters of T_i) is left. In the latter case we may add a vertex in the middle of the edge to create a unique root. The time required to determine a unique root in each tree is linear in V, the number of vertices in the trees, if a careful implementation is done.

The difficulty in determining tree isomorphism is that the edges at any vertex have no fixed order. If we could convert each tree into a canonical ordered tree, then testing tree isomorphism would be easy; we merely test for equality of the canonical ordered trees. Thus we need an algorithm which orders the edges at each vertex of the tree. Several authors [Busacker and Saaty (1965A), Lederberg (1964A), Scions (1968A), Weinberg (1965A)] present ordering methods, all virtually the same. Edmond's algorithm [Busacker and Saaty (1965A)] is a good example. Given a rooted tree, the vertices at each level are ordered, starting with those farthest away from the root. After the vertices at level i are ordered, to each vertex at level $i - 1$ is attached a list of its sons (the adjacent vertices at level i). The vertices at level $i - 1$ are then ordered lexicographically on the lists, according to the order already assigned to the vertices at level i. Once the vertices at each level are ordered, a canonical tree is easy to construct. Neither Edmonds nor any of the other authors who describe this technique note that the ordering process is tricky and must be done carefully if an $O(V)$ time bound is to be achieved.

It is useful to generalize the tree isomorphism problem slightly. We shall allow a set of labels to be attached to each vertex. Each label ℓ must be in the range $1 \leq \ell \leq V$, if V is the number of vertices in the tree. Two labelled trees are isomorphic if they may be matched as unlabelled trees and if any two matched vertices have identical label sets. We shall describe an isomorphism algorithm for labelled trees which requires $O(V,L)$

time, if V is the number of vertices and L the total size of the label sets. All sorting will be done using radix, or bucket sorting, using $2V + 1$ or fewer buckets. Thus the algorithm is suitable for implementation on a random-access computer.

If the trees are unrooted, unique roots are found using the method described above. Level numbers are calculated for all vertices. (The root is level 0.) Next, for each occurrence of a label ℓ an ordered pair (i,ℓ) is constructed; i is the level of this occurrence of ℓ. This set of ordered pairs is sorted lexicographically using two radix sorts to give, for each level i, a list $\mathcal{L}_i$ of the labels (in order) occurring at this level.

Next, we apply Edmond's algorithm, starting at the highest level vertices and working toward the root. Let k be the current level. The vertices at level $k + 1$ will already have been ordered and assigned numbers from 1 to N_{k+1}, where $N_{k+1} \leq V_{k+1}$ and V_{k+1} is the number of vertices at level $k + 1$. Using the list $\mathcal{L}_k$, the labels at level k are changed. The lowest is changed to $N_{k+1} + 1$, the next lowest to $N_{k+1} + 2$, and so on. Next, for each vertex v at level k, an index list containing the numbers of all its sons and all its labels is constructed. The numbers in these lists will be in order if the lists are constructed in the following way: For each son numbered i, an entry is made in its father's index list. This step is repeated for each i in the range $1 \leq i \leq N_{k+1}$. The label numbers are then entered similarly. The sons (the vertices at level $k + 1$) will be in order because of the processing done at level $k + 1$.

Now the vertices at level k must be ordered lexicographically on their index lists. Each number n occurring in an index list is converted into an ordered pair (i,n); i is the position of n in its index list. The set of pairs P is sorted lexicographically using two radix sorts to give a list P_i, for each position i, of the numbers (in order) occurring at that position in the index lists.

Let m be the length of the longest index list. We use m radix sorts, each with $N_{k+1} + |\mathcal{L}_k|$ buckets, to order the index lists lexicographically. During the ith pass, we add to the partially sorted set S of vertices all those whose index list has length $m - i + 1$. We then place vertex $v \in S$ into the buckets according to the value of the $m - i + 1$ th number in the index list of v. The non-empty buckets are emptied in order by referring to the list P_{m-i+1}.

After m passes, the vertices at level k are ordered lexicographically on their index lists. These vertices are numbered from 1 to N_k for some $N_k \leq V_k$; two vertices receive the same number if their index lists are the same. This step completes the processing for level k . After the vertices at each level are ordered, it is easy to construct a canonical ordered tree for each tree, and testing isomorphism is then a simple equality test.

Finding a root for each tree requires O(V) time. Constructing the label lists $\mathcal{L}_i$ requires O(L) time. Constructing the index lists at level k requires $O(V_{k+1} + |\mathcal{L}_k|)$ time. Ordering the vertices at level k lexicographically on their index lists requires m passes but only $O(V_{k+1} + |\mathcal{L}_k|)$ time, because only nonempty buckets are emptied on each pass and the total number of entries made in the buckets is $V_{k+1} + |\mathcal{L}_k|$. Since $\sum_k (V_k + |\mathcal{L}_k|) \leq V + L$, the entire algorithm has an O(V,L) time bound. If the trees are unlabelled, or if each tree has at most one label, the tree isomorphism algorithm requires O(V) time.

ISOMORPHISM OF TRIPLY CONNECTED PLANAR GRAPHS

In this section we describe an algorithm for partitioning a set of triply connected planar graphs into subsets of isomorphic graphs. The asymptotic running time of the algorithm grows as $V \log V$ where V is the total number of vertices in all graphs. An algorithm for such a partitioning [Hopcroft (1971)] based on converting the graphs to finite automata was previously used. However, working directly with graphs leads to major simplifications since the finite automata which are obtained from conversion of planar graphs are a very restricted subset of all finite automata and the full power to partition arbitrary finite automata is not needed.

Let G be a planar graph whose connected components are triply connected pieces other than n-gons or n-bonds. Consider a fixed embedding of G in the plane. We treat each edge of G as two directed edges. Let (v_1,v_2) be a directed edge. We denote (v_2,v_1) by $(v_1,v_2)^r$. We write $(v_1,v_2) \overset{R}{\vdash} (v_2,v_3)$ and $(v_1,v_2) \overset{L}{\vdash} (v_2,v_4)$ where (v_2,v_3) and (v_2,v_4) are edges bounding the faces to the right and left, respectively, of (v_1,v_2).

Let ε denote the string of length zero. For each edge e we write $e \overset{\varepsilon}{\vdash} e$. Let x be a string consisting of R's and L's. If $e_1 \overset{x}{\vdash} e_2$ and $e_2 \overset{R}{\vdash} e_3$ (or $e_2 \overset{L}{\vdash} e_3$) we write $e_1 \overset{xR}{\vdash} e_3$ (or $e_1 \overset{xL}{\vdash} e_3$). We write $e_1 \vdash e_2$ if x is understood. Intuitively we write $e_1 \overset{x}{\vdash} e_2$ if e_2 is reached by starting at e_1 and and traversing a path in the graph dictated by x. Each symbol of x dictates which way to turn on entering a vertex. On entering a vertex by edge e, leave the vertex by the edge immediately to the right or left of the edge e depending on whether the corresponding symbol of x is R or L respectively. Note that $e_1 \overset{*}{\Rightarrow} e_2$ does not necessarily imply $e_1 \vdash e_2$ since $e_1 \overset{*}{\Rightarrow} e_2$ denotes an arbitrary path and $e_1 \vdash e_2$ denotes a special type of path. Let λ be a mapping of edges into the integers such that $\lambda(e_1) = \lambda(e_2)$ if and only if the number of edges on the face to the right (left) of e_1 is the same as the number of edges on the face to the right (left) of e_2, the degrees of the heads of e_1 and e_2 are the same, and the degrees of the tails of e_1 and e_2 are the same.

Lemma 5. Let e be a directed edge and let v be a vertex in the connected component containing e.

1) There exists an edge e_1 directed into v such that $e \vdash e_1$.
2) Let e_1 be a directed edge into v. If v is of odd degree $e \vdash e_1$. If v is of even degree either $e \vdash e_1$ or $e \vdash e_1^r$.

Proof. (1) It suffices to show that for any v adjacent to the head of e there exists an edge e' directed into v such that $e \vdash e'$. Let $e_1, e_2, \ldots, e_m$ be the edges directed into the head of e. Consider the path e_1, e_2^r, followed by edges around the face to the left of e_2^r until e_3, e_4^r, followed by edges around the face to the left of e_4 and so on back to e_1. This path enters every vertex adjacent to v.

(2) Follow path to v and then twice around v by a path similar to that in (1).

Lemma 6. Let e_1, e_2, e_3 and e_4 be directed edges.

1) If $e_1 \vdash e_2$ then $e_2 \vdash e_1$ and $e_1^r \vdash e_2^r$.
2) If $e_1 \vdash e_1^r$ then $e_1 \vdash e_2$.
3) Assume there exists a path $p_1\colon e_1 \overset{*}{\Rightarrow} e_3$. Further assume that there exists a corresponding path $p_2\colon e_2 \overset{*}{\Rightarrow} e_4$. By corresponding we mean that both paths are the same length, that the number of edges clockwise between the edge into and the edge out of corresponding vertices agree, and that corresponding edges have the same value of λ. Then either

   (a) $e_1 \vdash e_3$ and $e_2 \vdash e_4$ or

   (b) $e_1 \vdash e_3^r$ and $e_2 \vdash e_4^r$ by the same sequence of right and left turns.

Proof. (1) Consider the path p from e_1 to e_2. Let e be the next-to-last edge in p. Without loss of generality we can assume e_2 bounds the face to the right of e and we need only show $e_2 \vdash e$. Clearly $e_2 \vdash e$ by a path which travels clockwise around the face to the right of e_2. Having established that $e_2 \vdash e_1$ it immediately follows that $e_1^r \vdash e_2^r$ by reversing all edges on the path $e_2 \vdash e_1$.

(2) By Lemma 5 either $e_1 \vdash e_2$ or $e_1 \vdash e_2^r$. If $e_1 \vdash e_2^r$ then by part 1 of Lemma 6 $e_1^r \vdash e_2$.

(3) By the construction in Lemma 5 either $e_1 \vdash e_3$ or $e_1 \vdash e_3^r$ by a path only using edges bounding faces adjacent to p_1 . A corresponding construction using p_2 yields the desired result.

Two edges e_1 and e_2 are said to be <u>distinguishable</u> if and only if there exists a string x such that $e_1 \overset{x}{\vdash} e_3$, $e_2 \overset{x}{\vdash} e_4$ and $\lambda(e_3) \neq \lambda(e_4)$. If e_1 and e_2 are not distinguishable they are said to be <u>indistinguishable</u>.

We need the following technical lemma.

<u>Lemma 7</u>. Let G be a biconnected planar graph. Let $(v_1,v_2)(v_2,v_3),\ldots,(v_{n-1},v_n)$ be a simple path p in G . Then there exists a face having an edge in common with the path which has the property that the set of all edges common to both the face and the path form a continuous segment of the path. Furthermore, when traversing an edge of the face while going from v_1 to v_n along the path, the face will be on the right.

<u>Proof</u>. See [Hopcroft (1971A)].

<u>Theorem 8</u>. Edges e and e' are indistinguishable if and only if there exists an isomorphism of the embedded version of G which maps e onto e' .

<u>Proof</u>. The if portion of the theorem is obvious. Namely, if e is mapped to e' by some isomorphism, then it is easily seen that e and e' are indistinguishable. The only if portion is more difficult to prove and we first establish it for regular degree three graphs.

Let G be regular of degree 3 and assume that edges e and e' are indistinguishable. We will now exhibit a method of constructing an isomorphism identifying e and e' . If e and e' are in the same connected component, then each edge not in the component is mapped to itself. If e and e' are in different components, say C_1 and C_2 , then each edge not in C_1 or C_2 is mapped to itself.

Identify edge e with e'. When two edges are identified, their reversals and their corresponding heads and tails are automatically identified. Whenever an edge e_1 is identified with an edge e_2, identify edges e_3 and e_4 where $e_1 \overset{R}{\vdash} e_3$ and $e_2 \overset{R}{\vdash} e_4$. If e_3 has already been identified with e_4, then select some pair of edges e_5 and e_6 which have already been identified, while e_7 and e_8 have not, where $e_5 \overset{L}{\vdash} e_7$ and $e_6 \overset{L}{\vdash} e_8$. Identify e_7 and e_8 and repeat the process. In other words always use the symbol R to obtain new edges, if possible, otherwise use L. This means that we will always identify edges along a path until we reach a pair of vertices already identified.

By Lemma 5(part 2), this procedure will yield the desired isomorphism unless a conflict arises. A conflict arises when we try to identify a vertex v_1 with a vertex v_2 which has already been identified with some $v_3 \neq v_1$. We now prove that such a situation is impossible.

Assume a conflict arises and consider the first such instance. One of the edges in the last pair identified must have completed a cycle. It is important to note that an edge that completes a cycle must terminate at a vertex which has both of the other edges already identified. The corresponding edge either did not complete a cycle (i.e. it terminated at an unidentified vertex) or it completed a different cycle (the end vertices of the two edges had previously been identified but not with each other). In the latter case the cycles are of different lengths. If both edges completed cycles, let c be the shorter of the two cycles. If only one cycle is completed, let c be that cycle. Let p be the path in the other graph corresponding to the vertices on the cycle c. The first and last vertices of p correspond to the same vertex c.

Since there is a cycle which is mapped to a simple path, select that cycle c which would map to a simple path but for which no cycle other than c containing only vertices from c and its interior would map to a simple path. By Lemma 7 some face is adjacent to p on the right and all edges of the face which are common to p form a continuous segment of p. Start identifying the edges around this face with edges on the interior of c. One of three cases occurs. (1) We return to a vertex on p before returning to a vertex on c, (2) We return to a vertex on c before returning to a vertex on p, or (3) Both events occur simultaneously. If (1) occurs, a face is identified with a non-closed path. This is impossible since λ contains information as to the number of edges around the face to the right or left of each edge. (2) is impossible since no cycle on the interior of c maps to a non-closed path. If (3) occurs we must have identified corresponding faces. This implies that the paths terminated at corresponding vertices and that c has been divided into two cycles c_1 and c_2. Assume c_1 corresponds to the face. Then cycle c_2 is mapped to a path, a contradiction. Since all possibilities lead to a contradiction, we are forced to conclude that no conflict can arise.

Having established the theorem for the special case where G is regular of degree 3, we now prove the theorem in general. Let G be an embedding of a planar graph all of whose connected components are triply connected. Assume that the edges e_1 and e_2 are indistinguishable but that there is no isomorphism mapping e_1 onto e_2. Let $\hat{G}$ be the graph obtained by expanding each vertex of degree $d > 3$ into a d-gon. Let $\hat{e}_1$ and $\hat{e}_2$ be the edges of $\hat{G}$ corresponding to e_1 and e_2. Since $\hat{G}$ is regular of degree 3, $\hat{e}_1$ and $\hat{e}_2$ must be distinguishable. Let $\hat{p}$ be the path that distinguishes $\hat{e}_1$ and $\hat{e}_2$. Clearly there is a corresponding path in G. By Lemma 6 (part 3) there exists an x which distinguishes e_1 and e_2, a contradiction.

The v log v isomorphism algorithm depends on an efficient algorithm for partitioning the edges of a planar graph into sets so that two edges are placed in the same set if and only if they are indistinguishable. This is done as follows.

Initially the edges are partitioned so that e_1 and e_2 are in the same set if and only if $\lambda(e_1) = \lambda(e_2)$. The index of each block of the partition except one is placed on a list called Rightlist and on a list called Leftlist. Let $B_1, B_2, \ldots, B_i$ be the current blocks of the partition. The blocks of the partition are refined by applying the following procedure until the Rightlist and Leftlist are empty.

Select the index j of some block from Rightlist or Leftlist and delete it from the list. Assume it came from Rightlist. For each e in B_j mark the edge e' defined by $e \overset{R}{\vdash} e'$. If block B_i contains both marked edges and unmarked edges partition it into two blocks B_{i_1} and B_{i_2} so that one contains only marked edges, the other only unmarked edges. Assume B_i is partitioned into blocks B_{i_1} and B_{i_2} where $|B_{i_1}| \leq |B_{i_2}|$. If the index i is already on Rightlist, replace it by i_1 and i_2 ; otherwise add only i_1 to Rightlist. Similarly if the index i is already on Leftlist, replace it by i_1 and i_2 ; otherwise add only i_1 to Leftlist.

Theorem 9. The above algorithm terminates and on termination two edges are in the same B_i if and only if they are indistinguishable.

Proof. (if) In the initialization phase edges are placed in different blocks only if they are immediately distinguishable. Subsequently a block is partitioned with e_1 and e_2 going into different blocks only if there exist e_3 and e_4 already in different blocks with $e_3 \overset{R}{\vdash} e_1$ and $e_4 \overset{R}{\vdash} e_2$ or $e_3 \overset{L}{\vdash} e_1$ and $e_4 \overset{L}{\vdash} e_2$. In the former case $e_1^r \overset{L}{\vdash} e_3^r$ and $e_2^r \overset{L}{\vdash} e_4^r$, and in the latter case $e_1^r \overset{R}{\vdash} e_3^r$ and $e_2^r \overset{L}{\vdash} e_4^r$. In either case e_1 and e_2 are distinguishable.

(only if) Assume e_1 and e_2 are distinguishable. We will prove that e_1 and e_2 are placed in different blocks of the partition by induction on the length n of the shortest string distinguishing e_1^r and e_2^r . Assume the induction hypothesis is true for sequences of length n and that the shortest sequence distinguishing e_1^r and e_2^r is of length n+1 . Without loss of generality assume the sequence is xR . Then x is of length n and distinguishes some e_3^r and e_4^r where $e_3 \overset{R}{\vdash} e_1$ and $e_4 \overset{R}{\vdash} e_2$. By the induction hypothesis e_3 and e_4 will be placed in separate blocks. The index of one of these blocks will go onto Rightlist and when it is removed e_1 and e_2 must be placed in separate blocks.

Theorem 10. The running time of the partitioning algorithm is bounded by kV log V for some k .

Proof. Clearly the running time of the algorithm is dominated by the time spent in partitioning blocks. Assume that at some point the blocks of the partition are $B_1, B_2, \ldots, B_m$. Let b_i be the cardinality of B_i . Let M be the set of integers from 1 to M , let I be the indices on Rightlist and let J be the indices on Leftlist. We claim that the time remaining is bounded by

$$T = k\left(\sum_{i\varepsilon I} b_i \log b_i + \sum_{i\varepsilon M-I} b_i \log (b_i/2) + \sum_{i\varepsilon J} b_i \log b_i + \sum_{i\varepsilon M-J} b_i \log(b_i/2)\right)$$

Clearly the bound holds when the algorithm has terminated. Consider what happens when an index i_0 is selected from I . Certain blocks will be partitioned. The remaining time will be bounded by some new T' . The time spent in partitioning is bounded by kb_{i_0} . We must show that

$$kb_{i_0} + T' \leq T .$$

Assume a block of size b_i is partitioned into blocks of size c_i and b_i-c_i where $c_i \leq b_i/2$. Clearly

$$b_i \log b_i \geq c_i \log c_i + (b_i-c_i)\log(b_i-c_i) \quad \text{and}$$

$$b_i \log b_i/2 \geq c_i \log c_i/2 + (b_i-c_i)\log(b_i-c_i)/2 .$$

Thus we need only show that

$$kb_{i_0} + kb_{i_0} \log(b_{i_0}/2) \leq k\, b_{i_0} \log b_{i_0} .$$ This follows since $b_{i_0} + b_{i_0} \log b_{i_0}/2 = b_{i_0}(1 + \log b_{i_0}/2) = b_{i_0} \log b_{i_0}$.

This completes the proof.

The partitioning algorithm can be used to partition a set of triply connected planar graphs into subsets of isomorphic graphs. Each triply connected planar graph has exactly two embeddings in the plane. One of the embeddings is obtained from the other by reversing the order of all edges around each vertex. We will refer to these embeddings as the clockwise and counterclockwise embeddings. Given a collection $G_1, G_2, \ldots, G_n$ of triply connected planar graphs, a composite graph G is formed consisting of two copies of each G_i . G is embedded in the plane so that one copy of each G_i has the clockwise embedding and one copy of each G_i has the counterclockwise embedding. The partitioning algorithm is applied to the embedding of G . As a consequence of

Theorems 8 and 9, G_i is isomorphic to G_j iff for any edge e in an embedding of G_i there exists an edge e' in one of the embeddings of G_j such that e and e' are in the same block of the partitioning of the edges. We can now easily partition the G_i into isomorphic graphs as follows: Place G_1 in the first block of the partition. Select an edge e of G_1. Scan the block of the edge partition containing e. For each j such that either embedding of G_j has an edge in the same block of the edge partition place G_j into the block containing G_1. Select the smallest k such that G_k is not already placed in Block 1 and place G_k in Block 2. Select an edge e of G_k and scan the block of the edge partition containing e as before. The whole process is repeated until every G_k is placed into some block.

The time necessary to complete the above process is bounded as follows: First the planar embeddings of the G_i must be determined. For each G_i the time required is proportional to the number of vertices in G_i [Tarjan (1972B)]. Thus the total time for this step is proportional to the number of vertices in the composite graph G. The time for the edge partitioning is bounded by $kE \log E$ where k is some constant and E is the number of edges in the composite graph G. Thus the total time is bounded by some constant times $E \log E$. Since in a planar graph $E \leq 3V-6$, the total time is $O(V \log V)$, where V is the number of vertices in G.

ISOMORPHISM ALGORITHM

Given two graphs G_1 and G_2, the isomorphism algorithm divides the graphs into connected components, subdivides each connected component into biconnected components and then further subdivides the biconnected components into triply connected components. The connectivity structure of each graph is represented as follows: Each graph is represented by a tree consisting of a root plus one vertex for each connected component. Each connected component is represented by a tree consisting of one vertex for each biconnected component and one vertex for each articulation point.

Let v_a be a vertex corresponding to an articulation point a and let v_B be a vertex corresponding to a biconnected component B containing a. Then v_a and v_B are connected by an edge. It is easy to see that the resulting graph is indeed a tree. The leaves of a tree representing a connected component are called <u>2-leaves</u>.

Each biconnected component B is also represented by a tree. The set of vertices consists of one vertex for each biarticulation point pair and one vertex for each triply connected component. If v_C represents a triply connected component C and v_{ab} represents a biarticulation point pair (a,b) then an edge connects v_C and v_{ab} if and only if a and b are both contained in C. The leaves of a tree representing a triply connected component are called <u>3-leaves</u>.

Consider the trees corresponding to connected components. Each 2-leaf is a biconnected component which corresponds to a tree-like structure of triply connected components. All triply connected components which are 3-leaves and are contained in the 2-leaves are assigned an ordered pair of numerical codes, such that equality of codes is equivalent to isomorphism. This is done by a method described below. These triply connected components are deleted and their codes are attached to new edges joining their biarticulation points in the remaining part of the graphs. This process creates new 3-leaves within the 2-leaves. These new 3-leaves are found and codes generated for them, and the process is repeated until each biconnected component which is a 2-leaf is reduced to a single edge. These edges are deleted, their codes being attached to the corresponding articulation points in the remaining part of the graphs. The new biconnected components which are 2-leaves are found and the process is repeated. Eventually, each connected component is reduced to a single vertex with an attached isomorphism code. The codes for the components of each graph are sorted and compared for equality. If they are equal, the graphs are isomorphic; if not, the graphs are non-isomorphic.

Consider any 3-leaf. Note that it has an orientation with respect to its biarticulation points. It is for this reason that we assign an ordered pair of numbers to the 3-leaf; the numbers are equal if and only if the 3-leaf is symmetric with respect to an exchange of the biarticulation points. In order to assign pairs of integers to a set of 3-leaves, the components are tested for planarity by a linear algorithm [Hopcroft and Tarjan (1972A), Tarjan (1972B)]. If any component is not planar, the isomorphism algorithm halts, since the entire graph is not planar. Assuming all 3-leaves are planar, the planarity algorithm constructs a planar representation of the component which is essentially unique since the component is triply connected. The $O(V \log V)$ algorithm described in Section 4 is used to determine the equivalence classes of isomorphic 3-leaves. Ad hoc integer codes are assigned to the biarticulation points of each 3-leaf according to the equivalence classes, in such a way that isomorphic components are assigned identical ordered pairs of integers and non-isomorphic components are assigned different ordered pairs.

The running time of the algorithm is dominated by the time required to partition the triply connected components into equivalence classes of isomorphic graphs. If V is the total number of vertices in each of the original graphs, then the partitioning requires $O(V \log V)$ time. The time to find biconnected and triply connected components, to test all 3-leaves for planarity, and to construct a planar representation, is $O(V)$; which is dominated by $O(V \log V)$.

EFFICIENCY OF EQUIVALENCE ALGORITHMS†

Michael J. Fischer

Massachusetts Institute of Technology
Cambridge, Massachusetts

1. INTRODUCTION

The equivalence problem is to determine the finest partition on a set that is consistent with a sequence of assertions of the form "$x \equiv y$". A strategy for doing this on a computer processes the assertions serially, maintaining always in storage a representation of the partition defined by the assertions so far encountered. To process the command "$x \equiv y$", the equivalence classes of x and y are determined. If they are the same, nothing further is done; otherwise the two classes are merged together.

Galler and Fischer (1964A) give an algorithm for solving this problem based on tree structures, and it also appears in Knuth (1968A). The items in each equivalence class are arranged in a tree, and each item except for the root contains a pointer to its father. The root contains a flag indicating that it is a root, and it may also contain other information relevant to the equivalence class as a whole.

Two operations are involved in processing a command "$x \equiv y$": first we must *find* the classes containing x and y, and then these classes are (possibly) *merged* together. The find is accomplished

---

†Work reported herein was conducted at the Artificial Intelligence Laboratory, a Massachusetts Institute of Technology research program supported in part by the Advanced Research Projects Agency of the Department of Defense and monitored by the Office of Naval Research under Contract Number N00014-70-A-0362 -0003.

by successively following the father links up the path from the given node until the root is encountered. To merge two trees together, the root of one is attached to the root of the other, and the former node is marked to indicate that it is no longer a root in the new data structure.

The time required to accomplish a find depends on the length of the path from the given node to the root of its tree, while the time to process a merge (given the roots of the two trees involved) is a constant. For definiteness, we let the cost of a merge be unity and the cost of a find be the number of nodes, including the endpoints, on the path from the given node to the root.

In this paper, we are interested in the way the cost of a sequence of instructions grows as a function of its length. Using the above algorithm, a sequence of n merge instructions can cause a tree to be built with a node v of depth n, so subsequent finds on that node will cost n+1 units each. The sequence consisting of the n merge instructions followed by n copies of a find(v) instruction will then cost n(n+2), and it is easy to see that $O(n^2)$ is an upper bound as well.

The above example suggests adding to the algorithm a "collapsing rule" which Knuth (1972A) attributes to Tritter. Every time a find instruction is executed, a second pass is made up the path from the given node to the root and each node on that path is attached directly to the root (except for the root itself). At worst this will only double the cost of the algorithm, and it may cause subsequent finds to be greatly speeded up. Indeed this turns out to be the case, for in Section 4 we show that the upper bound drops to $O(n^{3/2})$ using this heuristic.

Another heuristic, the "weighting rule", was studied by Hopcroft and Ullman (1971A) and previously known to several others. When performing a merge, an attempt is made to keep the trees balanced by always attaching the tree with the smaller number of nodes to the root of the tree with the larger number. To do this efficiently requires that extra storage be associated with each root in which to record the number of nodes in its tree. Hopcroft and Ullman show that with the weighting, a tree of n nodes can have height at most log n, and it follows that an instruction sequence of length n>1 can therefore have cost no greater than O(n log n). Moreover, in the absence of the collapsing rule, it is easy to construct instruction sequences whose cost does grow as n log n.

Combining the collapsing rule with the weighting rule yields an algorithm superior to those using either heuristic alone. With only the collapsing rule, we exhibit in Section 3 sequences whose cost grows proportionally to n log n, where n is the length of the sequence, and as we remarked above, a similar lower bound holds for just the weighting rule alone. Combining both heuristics, we derive in Section 4 an O(n log log n) upper bound. Hopcroft and Ullman (1971A) claim that the upper bound is actually linear. However, we have a counterexample to one of their earlier lemmas, and although this difficulty can be overcome, we are unable to follow the final part of their argument.

2. THE ALGORITHMS

An *equivalence program* over the set E is any sequence of instructions of the form find(a) where a is an element of E, or merge(A,B,C) where A, B and C are names of equivalence classes. (Cf. Hopcroft and Ullman (1971A).) Find(a) returns the name of the equivalence class of which a is a member, and merge(A,B,C) combines classes A and B into a single new class C.

We now consider two algorithms which can be used to implement equivalence programs. We first need some notation.

A *forest* F is a set of oriented (unordered) trees over some set V(F) of nodes. If v is a node, then *depth*[F](v) is the length of the path in F from v to a root, and *height*[F](v) is the maximum length of a path in F from v to a leaf. The depth and height will be written simply *depth*(v) and *height*(v) when the forest F is understood. The height of a tree A, *height*(A), is the height of its root.

The algorithms are built from three kinds of instructions which operate on a forest F. If v is a node, then *find*(v) does the following:

1. If v is a root, or if father(v) is a root, then F is left unchanged.
2. Otherwise, let $v=v_0,v_1,\ldots,v_k$ be the (unique) path from v to the root v_k. Then F is modified by making v_k the father of each of the nodes $v_0,\ldots,v_{k-2}$.

The cost of find(v) is 1 + depth(v).

The instruction *U-merge*(u,v) has unit cost and is defined only when u and v are both roots. It causes the node u to become a direct descendant of v (and hence u is no longer a root).

For any node v, let *weight*(v) be the number of nodes in the subtree rooted by v (and including v itself). The instruction *W-merge*(u,v) also has unit cost and is defined only when u and v are both roots. If weight(u) $\leq$ weight(v), it behaves exactly like U-merge(u,v); otherwise, it causes the node v to become a direct descendant of u.

We define a *U-program* to be any sequence of instructions consisting solely of finds and U-merges. Similarly, a *W-program* is any sequence of finds and W-merges.

Let α be a U- (W-)program. Then $T(\alpha)$ is the total cost of executing the instructions of α in sequence, starting from an initial forest F_0 in which every node is a root. $T(\alpha)$ is undefined if any of the instructions in α is undefined.

3. A LOWER BOUND FOR THE COST OF THE UNWEIGHTED ALGORITHM

In this section, we show how to find, for each $n > 0$, a U-program α of length n such that $T(\alpha) > cn(\log n)$ for some constant c independent of n.

We begin by defining inductively for each n a class S_n of trees:

(i) Any tree consisting of just a single node is an S_0 tree.

(ii) Let A and B be S_{n-1} trees, and assume that A and B have no nodes in common. Then the tree obtained by attaching the root of A to the root of B is an S_n tree.

Figure 3.1 illustrates the building of an S_n tree, and Figure 3.2 shows an S_4 tree.

Lemma 3.1. Let A be an S_n tree. Then A has 2^n nodes, height(A) = n, and A contains a unique node of depth n.

Proof. Trivial induction on n. □

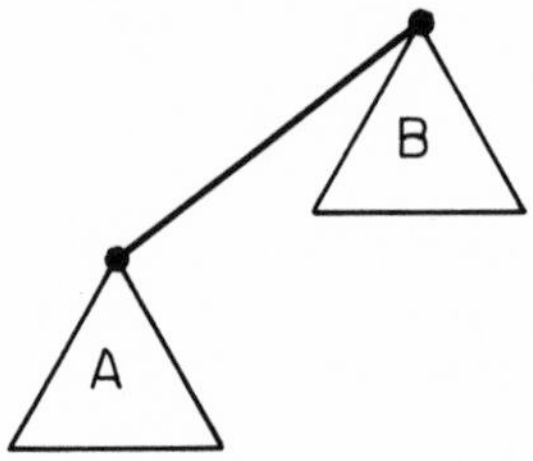

Figure 3.1. Definition of an S_n tree.

In light of the lemma, we define the *handle* of an S_n tree to be the unique node of depth n.

Two alternate characterizations of S_n trees are illustrated in Figure 3.3 and stated in:

Lemma 3.2. Let A be an S_n tree with handle v.

(a) There exist disjoint trees $A_0, \ldots, A_{n-1}$ not containing v with roots $a_0, \ldots, a_{n-1}$ respectively such that (1) A_i is an S_i tree, $0 \leq i \leq n-1$, and (2) A is the result of attaching v to a_0 and a_i to a_{i+1} for each i, $0 \leq i < n-1$.

(b) There exist disjoint trees $A'_0, \ldots, A'_{n-1}$ with roots $a'_0, \ldots, a'_{n-1}$ respectively and a node u not in any A'_i such that (1) A'_i is an S_i tree, $0 \leq i \leq n-1$, and (2) A is the result of attaching a'_i to u for each i, $0 \leq i \leq n-1$. Moreover, v is the handle of A'_{n-1}.

Proof. Again the proof is a trivial induction on n and is omitted. □

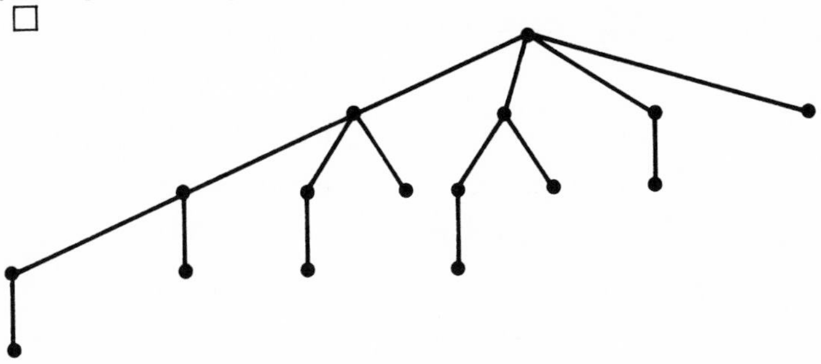

Figure 3.2. An S_4 tree.

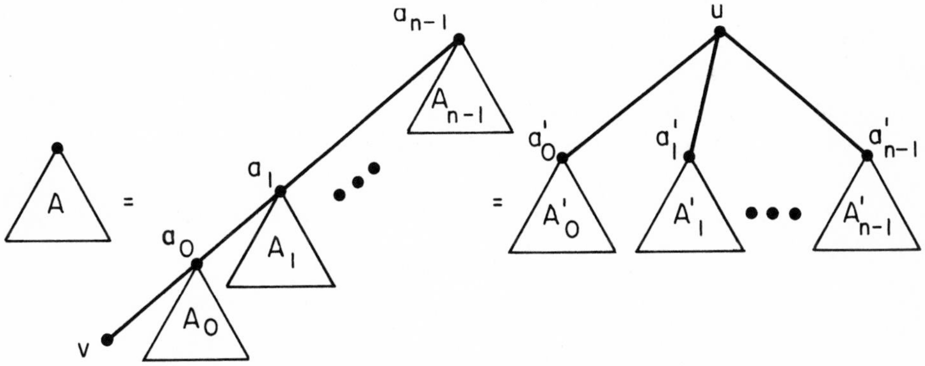

Figure 3.3. Decompositions of an S_n tree A.

The remarkable property of an S_n tree is that it is self-reproducing in the sense that if an S_n tree A is embedded in a larger tree B so that the root of A has depth > 0 in B, then a find on the handle of A (which collapses the path above the handle) costs at least n+2 and the resulting tree still has an S_n tree embedded in it!

We now make these notions more precise.

Definition. Let A and B be trees. A one-one function η: $V(A) \to V(B)$ is an *embedding* of A in B if for all $u,v \in V(A)$, $u = \text{father}(v)$ iff $\eta(u) = \text{father}(\eta(v))$. η is *initial* (*proper*) if η maps (does not map) the root of A onto the root of B. We say that A is *initially* (*properly*) *embeddable* in B if there exists an initial (proper) embedding of A in B.

Lemma 3.3. Let A be an S_n tree with handle v, and assume η is a proper embedding of A in a tree P. Then A' is initially embeddable in the tree P', where A' is an S_n tree and P' results from the instruction find(η(v)) on P.

Proof. The trees described below are illustrated in Figure 3.4.

Let A be an S_n tree with handle v, and assume η is a proper embedding of A in P. By Lemma 3.2(a), we may assume that $v, a_0, \ldots, a_{n-1}$ is the path from v to the root of A, and $a_0, \ldots, a_{n-1}$ are the roots of disjoint subtrees $A_0, \ldots, A_{n-1}$ respectively, where each A_i is an S_i tree, $0 \le i \le n-1$.

For each i, $0 \le i \le n-1$, let P_i be the subtree of P consisting of the nodes in $\{\eta(u) \mid u \in V(A_i)\}$.

Let A' be the tree formed as in Lemma 3.2(b) by linking each of the nodes a_i to a new node a'. Then A' is an S_n tree.

Let P' result from the execution of the instruction find(η(v)) on P, and let ρ be the root of P'.

Finally, define a mapping η' from the nodes of A' to the nodes of P':

$$\eta'(u) = \begin{cases} \eta(u) & \text{if } u \in V(A_i) \text{ for some } i,\ 0 \le i \le n-1; \\ \rho & \text{if } u = a'. \end{cases}$$

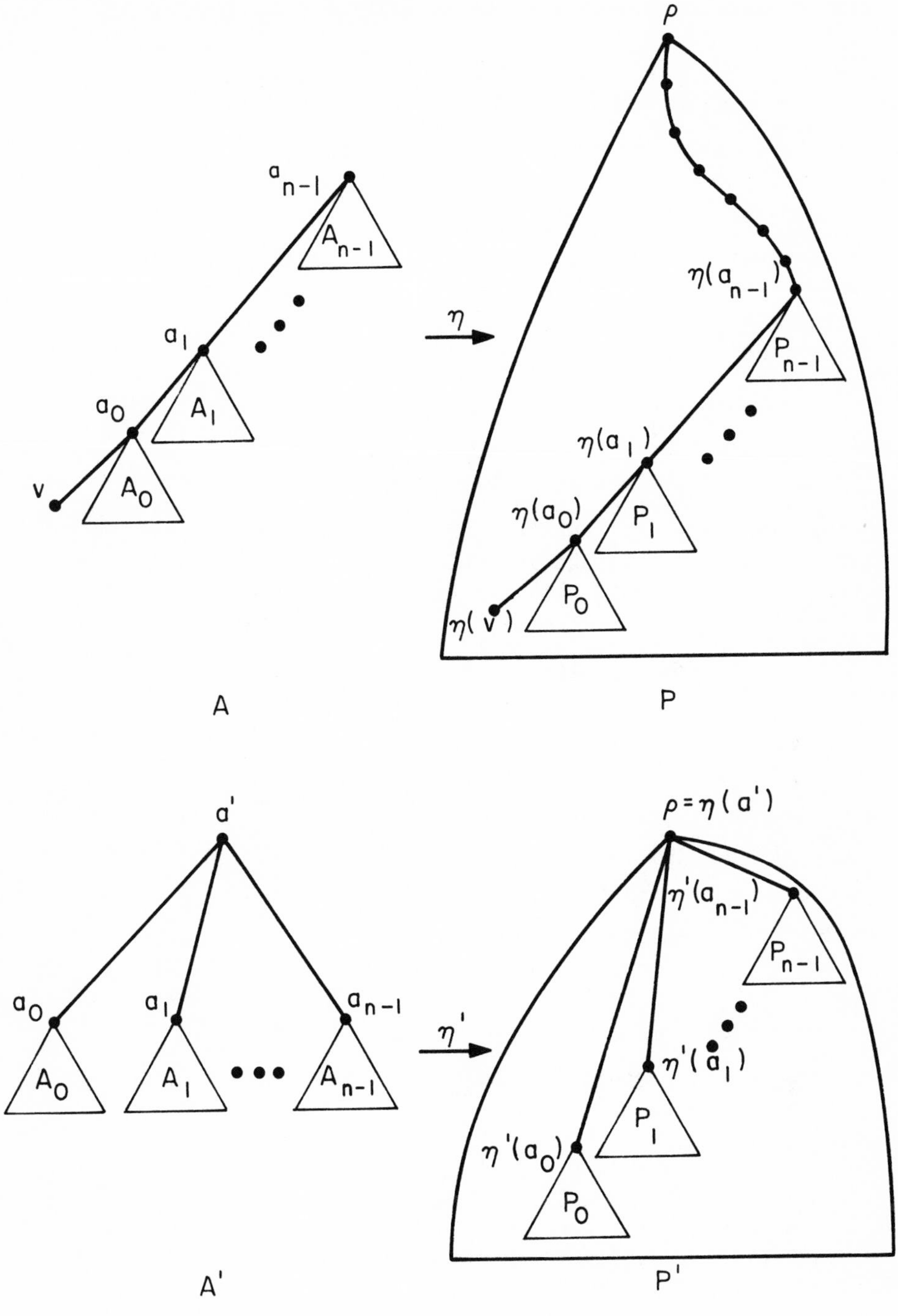

Figure 3.4. Trees in the proof of Lemma 3.3.

It remains to show that η' is an initial embedding of A' in P'

Let π be the path from $\eta(v)$ to the root of P. From the definition of embedding, each of the nodes $\eta(v)$, $\eta(a_0),\ldots,\eta(a_{n-1})$ appears on π, and no node in P_i except for $\eta(a_i)$ is in π, $0 \le i \le n-1$.

As a consequence of the find, each of the nodes $\eta(a_i)$ is linked directly to the root ρ of P', and since the path π did not run through any nodes of P_i except for the root, P_i is a subtree of P' linked directly to ρ. It is easily verified that η' is an initial embedding of A' in P'. □

We now construct a costly U-program. First build an S_k tree. Then alternately "push" it down by merging it to a new node, and perform a find on the handle. This find costs k+2 units and it leaves us with a new tree in which an S_k tree is initially embedded. Thus we can repeat the "merge, find" sequence as often as we wish, yielding an average instruction time that approaches (k+3)/2. Since we can do this for arbitrary k, the cost of U-programs cannot be linear in their length. In fact, we show:

Theorem 1. For any n>0, there exists a U-program α of length n such that $T(\alpha) > cn(\log n)$ for some constant c independent of n.

Proof. Let $a_1,a_2,\ldots$ be a sequence of distinct nodes, and let β be a program of 2^k-1 U-merges which builds an S_k tree out of the nodes $a_1,\ldots,a_{2^k}$. For each $i \ge 1$, let v_i be the handle and r_i the root of the tree that results from the sequence $\beta,\gamma_1,\ldots,\gamma_{i-1}$ and define γ_i = "U-merge(r_i, a_{2^k+i}), find(v_i)". Let α be the sequence $\beta,\gamma_1,\ldots,\gamma_m$, where $m = 2^k-1$. Then $T(\alpha) = (2^k-1) + m(k+3)$, and the length of α is $n = 3m$, so

$$T(\alpha) = \frac{n}{3} + \frac{n(k+3)}{3} > cn(\log n) \qquad (3.1)$$

for some constant c.

For n not of the form $3(2^k-1)$, we form the next shorter sequence that is of that form and then extend it arbitrarily to get a sequence of length exactly n. This will have the effect only of changing the constant in (3.1). □

4. UPPER BOUNDS

We get upper bounds on the two algorithms by considering a slight generalization of a find instruction. Find(u,v) behaves like a find(u) where we pretend that v is the root. More precisely, *find*(u,v) is defined only if v is an ancestor of u. If that is the case, let $u=u_0,u_1,\ldots,u_k=v$ be the path from u to v. Then find(u,v) causes each of the nodes $u_0,\ldots,u_{k-2}$ to be attached directly to v. Its cost is defined to be k+1. A sequence of generalized find and U- (W-)merge instructions is called a *generalized U- (W-)program.*

Notation. Let F be a forest and α a program. Then F:α is the forest that results from F by executing the instructions in α.

Lemma 4.1. Let u be any node in a forest F. Then there exists a node v in F such that F:find(u) = F:find(u,v) and the costs of executing find(u) and find(u,v) are the same.

Proof. Choose v to be the root of the tree containing u. □

Applying Lemma 4.1 in turn to each of the find instructions in a U- or W-program α gives the following:

Lemma 4.2. Let α be a U- (W-)program and F a forest. Then there exists a generalized U- (W-)program β such that F:α = F:β and $T(\alpha) = T(\beta)$.

Generalized programs are convenient to deal with because there is no loss of generality in restricting attention to programs in which all the merges precede all the finds.

Lemma 4.3. Let F be a forest containing the nodes p, q, u and v and let M be the instruction U-merge(p,q) (W-merge(p,q)). Let α_1 = "find(u,v), M" and α_2 = "M, find(u,v)". If α_1 is defined on F, then F:α_1 = F:α_2 and $T(\alpha_1) = T(\alpha_2)$.

Proof. The only possible effects of M are to change the father of p to be q, or to change the father of q to be p. Similarly, the only possible effects of the instruction find(u,v) are to change the fathers of the nodes on the path from u to v (but not including the last two such nodes). Since α_1 is defined, then v is an ancestor of u and both p and q are roots in F; hence the sets of father links changed by the two instructions are disjoint. Moreover, the choice of whether to link p to q or q to p in case M is a W-merge instruction depends only on the weights of p and q, and the weight of a root is not affected by a find instruction. Hence, neither instruction affects the action of the other, so $F:\alpha_1 = F:\alpha_2$ and $T(\alpha_1) = T(\alpha_2)$. □

Lemma 4.3 enables one to convert a generalized program into an equivalent one in which all the merges precede all the finds.

Lemma 4.4. Let α be a generalized program, and let β result from α by moving all the merge instructions left in the sequence before all the finds, but preserving the order of the merges and the order of the finds. Then $F:\alpha = F:\beta$ and $T(\alpha) = T(\beta)$.

To bound the cost of a generalized U-program, we consider the effects of a U-merge and a generalized find instruction on the *total path length* of a forest F, defined to be

$$\sum_{v \in V(F)} \text{depth}(v).$$

Lemma 4.5. Let α be a sequence of n U-merge instructions and let $F = F_0:\alpha$. Then the total path length of $F \leq n^2$.

Proof. No node in F can have depth > n, and at most n nodes have non-zero depth. Hence, the total path length $\leq n^2$. □

Lemma 4.6. A generalized find instruction of cost $\ell > 2$ reduces the total path length by at least $(\ell-2)^2/2$.

Proof. Let find(u,v) be an instruction of cost ℓ. Then there is a path $u=u_0,u_1,\ldots,u_{\ell-1}=v$ from u to v. For each i, $0 \leq i \leq \ell-3$, the find causes the depth of node u_i to become one plus the depth of v, so the reduction in total path length is at least

$$\sum_{i=0}^{\ell-3} (\text{depth}(u_i) - (1+\text{depth}(v))) = \sum_{i=0}^{\ell-3} (\ell-2-i) = \sum_{j=1}^{\ell-2} j \geq \frac{(\ell-2)^2}{2}. \square$$

Theorem 2. Let α be a U-program of length n. Then $T(\alpha) < cn^{3/2}$ for some constant c independent of n.

Proof. By Lemma 4.2, it suffices to bound a generalized U-program α instead, and by Lemma 4.4, we may assume that all the U-merges in α precede all the finds.

A program of length n clearly has at most n merge instructions and at most n find instructions. Let ℓ_i be the cost of the ith find instruction if there is one and 0 if not. Clearly,

$$T(\alpha) \leq n + \sum_{i=1}^{n} \ell_i. \tag{4.1}$$

By Lemma 4.5, the forest after executing the merge instructions in α can have a total path length of at most n^2. Only the find instructions of cost greater than two affect the tree, so let $I = \{i \mid \ell_i > 2\}$. If $i \in I$, Lemma 4.6 asserts that the ith find instruction decreases the total path length by at least $(\ell_i - 2)^2/2$. The total path length at the end cannot be negative, so

$$n^2 \geq \frac{1}{2} \sum_{i \in I} (\ell_i - 2)^2 \geq \frac{1}{2} \sum_{i=1}^{n} (\ell_i - 2)^2 - 2n \tag{4.2}$$

or

$$6n^2 \geq \sum_{i=1}^{n} (\ell_i - 2)^2. \tag{4.3}$$

The maximum value for $\sum_{i=1}^{n} \ell_i$ is achieved when all the ℓ_i's are equal, for if they are not all the same, replacing each by the mean ℓ can only cause $\sum_{i=1}^{n} (\ell_i - 2)^2$ to decrease. Hence, from (4.1) and (4.3) we get

$$T(\alpha) \leq n + n\ell \tag{4.4}$$

where ℓ is subject to the constraint that

$$6n^2 \geq n(\ell - 2)^2. \tag{4.5}$$

From (4.5),

$$\ell \leq 2 + \sqrt{6n} \tag{4.6}$$

and substituting into (4.4), we get

$$T(\alpha) \leq n + n(2 + \sqrt{6n}) < 6n^{3/2}. \quad \square \tag{4.7}$$

For the case of the weighted algorithm, we prove an upper bound of $O(n \log \log n)$ using a method similar to our proof of Theorem 2.

We say that a forest F is *buildable* if it can be obtained from F_0 by a sequence of W-merge instructions. Buildable forests have the important property that most nodes have low height.

Lemma 4.7 (Hopcroft and Ullman (1971A)). Let F be a buildable forest. If v is a node in F of height h, then weight(v) $\geq 2^h$.

Proof. The result follows readily by induction on h. We leave the details to the reader. □

Corollary. Let α be a sequence of W-merge instructions of length n and let $F = F_0{:}\alpha$. For any $h \geq 0$, F contains at most $n/2^h$ non-roots of height h.

Proof. F has exactly n non-roots, for each W-merge changes one root to a non-root. Suppose $u_1,\ldots,u_k$ are non-roots of height h. By the lemma, weight$(u_i) \geq 2^h$, and all the nodes counted in the weight of u_i are non-roots, $1 \leq i \leq k$. Hence,

$$n \geq \sum_{i=1}^{k} \text{weight}(u_i) \geq k \cdot 2^h, \tag{4.8}$$

so $k \leq n/2^h$.

Instead of looking at total path length, we consider a quantity Q(F,G) which depends on two forests F and G. Our interest is in the case where F is a buildable forest and G results from F by a sequence of generalized finds, although our definition applies whenever V(F) = V(G):

$$Q(F,G) = \sum_{v \in V(F)} \text{depth}[G](v) \cdot 2^{\text{height}[F](v)}. \quad \square \tag{4.9}$$

Lemma 4.8. Let α be a sequence of W-merge instructions of length $n \geq 1$ and let $F = F_0{:}\alpha$. Then $Q(F,F) \leq n(\log(n+1))^2$.

Proof. No tree in F can have more than n+1 nodes. By Lemma 4.7, a root can have height at most log(n+1), so no node has height or depth greater than log(n+1).

Let $N = \{\, v \in V(F) \mid \text{depth}[F](v) > 0 \,\}$ be the set of non-roots of F. From (4.9), we get

$$Q(F,F) \leq \log(n+1) \cdot \sum_{v \in N} 2^{\text{height}[F](v)}. \tag{4.10}$$

We now wish to bound $R(F) = \sum_{v \in N} 2^{\text{height}[F](v)}$. Since a root has height at most log(n+1), any node $v \in N$ has height at most $H = \log(n+1) - 1$, so summing over the heights of nodes,

$$R(F) = \sum_{h=0}^{\lfloor H \rfloor} (\#\text{ nodes in N of height h})\cdot 2^h \tag{4.11}$$

By the corollary to Lemma 4.7, the number of nodes in N of height h is at most $n/2^h$, so

$$R(F) \le \sum_{h=0}^{\lfloor H \rfloor} (\frac{n}{2^h})\cdot 2^h \le (H+1)n = n\cdot\log(n+1). \tag{4.12}$$

Substituting (4.12) into (4.10) gives the desired result. □

Lemma 4.9. Let F be a buildable forest, ϕ a sequence of generalized finds, and let $G = F{:}\phi$. If u is a descendant of v in G and $u \neq v$, then height[F](u) < height[F](v).

Proof. It is easy to show by induction on the length of ϕ that if u is a descendant of v in G, then u is also a descendant of v in F. By the definition of height, it follows that height[F](u) < height[F](v). □

Lemma 4.10. Let F be a buildable forest, ϕ a sequence of generalized finds, and let $G = F{:}\phi$. Assume find(u,v) is defined on G, has cost $\ell > 2$, and results in a forest G'. Then $Q(F,G) - Q(F,G') \ge 2^{\ell-3}$.

Proof. Let $u=u_0,\ldots,u_{\ell-1}=v$ be the path from u to v in G. By Lemma 4.9, the heights in F of the nodes in the path are monotone increasing, and since heights are integral, $\text{height}[F](u_{\ell-3}) \ge \ell-3$. The instruction find(u,v) does not increase the depth of any node and it decreases the depth of $u_{\ell-3}$ by one, so $Q(F,G) - Q(F,G') \ge 2^{\text{height}[F](u_{\ell-3})} \ge 2^{\ell-3}$. □

Theorem 3. Let α be a W-program of length $n \ge 4$. Then $T(\alpha) < cn(\log\log n)$ for some constant c independent of n.

Proof. By Lemmas 4.2 and 4.4, it suffices to prove the theorem for a generalized W-program $\alpha = \mu\phi$ of length n, where μ is a sequence of W-merge instructions and ϕ is a sequence of generalized find instructions.

The lengths of μ and ϕ are clearly both at most n. Let ℓ_i be the cost of the ith find instruction if there is one and 0 if not. Then

$$T(\alpha) \le n + \sum_{i=1}^{n} \ell_i. \tag{4.13}$$

Now, let $F = F_0 : \mu$. By Lemma 4.8,

$$Q(F,F) \leq n(\log(n+1))^2. \tag{4.14}$$

Only find instructions of cost greater than two affect the forest, so let $I = \{i \mid \ell_i > 2\}$ and let $G = F:\phi$. By repeated use of Lemma 4.10,

$$Q(F,F) - Q(F,G) \geq \sum_{i\varepsilon I} 2^{(\ell_i - 3)} \geq \left(\sum_{i=1}^{n} 2^{(\ell_i - 3)}\right) - n. \tag{4.15}$$

Since $Q(F,G) \geq 0$, we conclude from (4.14) and (4.15) that

$$n(\log(n+1))^2 \geq \sum_{i=1}^{n} 2^{(\ell_i - 3)} - n, \tag{4.16}$$

so

$$2n(\log(n+1))^2 \geq \sum_{i=1}^{n} 2^{(\ell_i - 3)}. \tag{4.17}$$

The maximum value for $\sum_{i=1}^{n} \ell_i$ is achieved when all the ℓ_i's are equal, for if they are not all the same, replacing each by the mean ℓ can only cause $\sum_{i=1}^{n} 2^{(\ell_i - 3)}$ to decrease. Hence, from (4.13) and (4.17), we get

$$T(\alpha) \leq n + n\ell \tag{4.18}$$

where ℓ is subject to the constraint that

$$2n(\log(n+1))^2 \geq n \cdot 2^{(\ell - 3)}. \tag{4.19}$$

Taking logarithms (to the base 2), we get

$$\ell \leq 3 + \log 2 + 2(\log\log(n+1)) \leq 6(\log\log(n+1)). \tag{4.20}$$

Substituting back into (4.18) yields

$$T(\alpha) \leq n + 6n(\log\log(n+1)) < 13n(\log\log n). \tag{4.21}$$

5. CONCLUSION

We have considered two heuristics, the collapsing rule and the weighting rule, which purportedly improve the basic tree-based equivalence algorithm. Our results, together with the remarks in the introduction, show that each heuristic does indeed improve the worst case behavior of the algorithm, and together they are better than either alone.

There is still a considerable gap between the lower and upper bounds we have been able to prove for the two algorithms employing the collapsing rule, and we are unable to show even that the weighted algorithm requires more than linear time. We leave as an open problem to construct any equivalence algorithm at all which can be proved to operate in linear time.

ACKNOWLEDGEMENT

The author wishes to express his appreciation to Albert Meyer for several enlightening discussions and to Patrick O'Neil for some ideas leading to a proof of Theorem 3. He is also grateful to John Hopcroft for pointing out an error in the original version of that proof.

PANEL DISCUSSION

The following is a slightly edited version of a tape of the panel discussion held at the end of the Symposium. The panel members were Charles M. Fiduccia, Robert Floyd, John E. Hopcroft, Richard M. Karp, Michael Paterson, Michael O. Rabin, Volker Strassen, and Shmuel Winograd. The moderator was Raymond E. Miller.

MILLER: I have asked the panelists to look over some general questions which I want to read to you. After I read them, I'll ask the panelists if they want to speak on any of these questions or make comments on any other matters of their choice. I am limiting, however, each of the panelists to five minutes during this period. Hopefully that won't take the full hour. After that time, we're going to open the panel for questions and discussion from the audience. In general, if the question is directed to a panelist, I would like the panelist to summarize the question before he answers it. I'll have to interrupt you if you don't, so please try to discipline yourselves.

There are four general questions that I think might be interesting. The first is: "How is the theory developing from originally being a scattering of a few results on lower bounds and some algorithms into a more unified theory? Are there general tools and techniques evolving both for developing algorithms and proving theorems about bounds, or is this still just a scattered field with no unity?"

The second question is: "What specific examples have been found to demonstrate how real computer computations were improved from studies of this type? That is, can you show numbers that say, 'Here, we used to do it this way, but now through this theory we have developed new techniques and we have saved this much time on a real program. Possibly the saving is some other measure than computation time.'"

The third question concerns the status of combinatorial type problems versus numerical problems. We've heard some of each during the Symposium. "Is the progress of the numerical type computations and understanding of them much ahead of the combinatorial? What future progress in each of these fields is expected?"

The last general question I have is: "Are there any important open problems that you wish to bring up to this audience?"

Gentlemen, who would like to have the floor?

WINOGRAD: I just got copies of the four questions about fifteen minutes ago and, therefore, I haven't given that much thought to them, so let me just start with some random thoughts that I have. I hope they will cover some of the questions.

I don't know how you people have felt -- I hadn't realized how much progress we had made, but sitting here for two and one-half days and hearing these talks helped to summarize it all for me. I'm rather optimistic. I think that a certain kind of a theory is developing. Concerning a unified approach, I think that the very nice paper by Volker Strassen, which will enable us to bring to bear some rather powerful algebraic techniques into solving those problems, is an indication of progress. It's an important way of showing that we are really getting to what some might call a unified theory rather than scattered results.

Another thing that was very heartening to me was to see the types of problems that people have looked at from a computational complexity point of view. There are the more algebraic ones and the more combinatorial ones like sorting and merging. We have started to get results on what I might call a more analytic problem. Maybe one day we will be able to look at a system of partial differential equations and be able to say something meaningful about how much work has to be involved in order to solve these types of problems. So the types of questions that we are looking at has increased tremendously. The combination of those two points, on the one hand the deepening of mathematical theory bringing to bear rather deep mathematical results, and on the other hand, the widening of the scope of the problems that we are looking at, is essentially what makes me really quite optimistic about this whole field.

STRASSEN: It seems to me that problems of computational complexity arise in almost any mathematical field and that techniques for deriving specific upper and lower bounds will depend on the field from which these problems are taken. For example, for polynomial evaluation, which is a problem from algebra, our tools are substitution of variables, transcendence degree, etc.; for the number theoretical problem of computing algebraic numbers, we have seen that Lionville's theorem is relevant; and for combinatorial problems, the natural tools are combinatorial. So I don't see complete unification, although within large areas of the theory, a great deal of unification can and will be achieved.

This conference has shown, in my opinion, that as of today it is mainly problems from algebra and combinatorics (including graph theory) that have led to a general and systematic investigation. The reason is, of course, that algebraic and combinatorial problems often are relatively simple. Whenever the concept of convergence is involved, the problems seem to be much more difficult. It is not surprising, therefore, that the present results in this area are not built upon general models like program or computation, but on more special schemes like iteration. I think, however, that in the long run computational complexity theory and classical numerical analysis will merge.

RABIN: Addressing myself first of all to the first question about the degree of unity within the field, I'm more optimistic than let's say Volker [Strassen], who spoke just before me. I think, from experience and observation, that the methods aren't really that diverse. Let me give you some examples. Let's take Floyd's lecture on transposing a bit matrix or rearranging records. Now, there is a close and clear relationship between the recursion that goes on there, and even in the way it's being programmed, and the finite Fourier transform, which was actually mentioned in the lecture itself. Let's take the problem of evaluation of polynomials and the various methods of preconditioning that are being used there, especially Mike Paterson's algorithms for parallel computation. There is a close relationship between this, almost an identical approach, and algorithms for fast addition of numbers in hardware. If you look, for example, at what Brent did, getting a very good hardware adder, then you see that the way the computation flows and the way you combine things, is very similar to what you do in evaluating polynomials in a situation of parallelism. And these examples could be multiplied.

In mathematics, there is a way to attack discrete problems like, for example, counting partitions, by analysis methods, such as generating functions and the classical theory of functions of a complex variable. I think that we'll find some of that approach also in this area of computational complexity. There are some results about sorting, for example, or how many steps are required to find the maximum of n numbers, which you can already do by analysis. It is also quite possible that some of the problems mentioned by Dick Karp in his lecture, even though they are 0-1 problems, may be approached through using real numbers and the ordinary methods of mathematics. In summation, if you look at it carefully, you can see common ideas which are a guide. It is true that in each particular discipline, you are going to get the best results by being very specific in that discipline.

With regard to the second question, there are some examples of actual applications of complexity theory to practical problems. Again, that was the situation where one was able to take ideas which come from preconditioning of polynomials and use them to actually speed up arithmetic in computers. So I'll just mention it without going into detail.

Now, important open problems. The thing which I would like most to see done is an answer to the questions raised by the work of Cook and by the results which Dick Karp presented yesterday. These mutual reductions, showing that various sets are complete, are marvelous. But I think that we are now forced into the position of saying which it is -- whether all of these problems are simultaneously of some polynomial size, or whether they are all exponential or superpolynomial, at least. There again you have an example of this commonality. The theory of mathematical machines, the type of questions about reduction from nondeterministic to deterministic, etc., was rather dying for a while, because people, and rightly, got tired and wanted to see new things. Now, you saw how the above problem can be answered by deciding something about equivalence between non-deterministic and deterministic machines. There is also another possible approach to these problems, not via machines, but again by analysis, linear, well, sort of linear algebra, but really harder methods.

An area which is entirely different, where, I think that things can be done and that those things that will be done might very well be of practical importance, is non-numerical problems of the type, say, that Bob [Floyd] was talking about. The re-arranging of records and sorting are obvious examples. But then you can also think about just set operations. How fast can you determine certain intersections between sets, or questions whether intersection of certain sets represented in a computer are empty or not. I don't want to go into detail, but only mention that for certain of these problems, if you take the Strassen matrix multiplication algorithm, it turns out that you can do them faster than by the ordinary known combinatorial methods. I am referring to this business of finding the transitive closure of a relation. So again, you have an example of commonality and I think it would be very nice to have set-algorithms for doing these problems, which are faster than the obvious ones. We know that they exist, and any such advance might also in certain situations of data processing represent a practical advance.

FLOYD: One of the questions we have here is: "How is theory developing from a scattering of lower bounds and algorithms to a unified theory?" I want to be the first courageous one to answer that, and reply, "Slowly." Let me go into a little more detail on another one of the questions: "What specific examples have been found to demonstrate how real computer computations were improved from studies of this type?" I think there are two aspects worth pointing out to that question. One is, I think computational complexity is playing a role something like the role that results in undecidability were playing perhaps ten years ago or so, not in making people better at what they could do, but rather telling them what they couldn't do and what they shouldn't bother trying to do. Around ten years ago, after the publication of the ALGOL 60 report and the discovery that the language was full of syntactic ambiguities, I was working for a while on trying to find an algorithm to test whether a formal language contained syntactic ambiguities or not. This question was eventually resolved when one of Bar-Hillel's papers on some decidability questions in context-free languages included some results which could be adapted to show that the ambiguity problem for context-free languages was undecidable. It simply saved me and some other people some time by keeping us from trying to do what we couldn't do; and, of course, a great many results in computational complexity have been of this form. They've been saying, "Well, don't bother trying to do things any better than you're already doing them. You can't."

The results that say that in principle you can do better than you are now doing don't seem to have had much effect on the actual conduct of computation yet. As for why that is, I think the reason is that in any particular approach to computational complexity, there is some definition of equivalent complexity involved, and there are different strengths of definition of equivalence. For example, in terms of most of the papers that we've heard at this conference on, say, root finding, one speaks of the degree of convergence of a method. This is an asymptotic property of the method. Within the current viewpoint, two methods are essentially considered equivalent if they have the same degree, so they are characterized basically by one real number representing an asymptotic property. This is a very weak equivalence relation; you can imagine two algorithms that would be equivalent under this measure, but would be extremely different before they went to the limit. I think this is true not only there, but in virtually any area we've mentioned. In the methods I was describing for permuting information, I was measuring the cost of an algorithm only by how many of these bounded size set unions were performed, whereas in actuality, one would have to worry about getting the information to the right places. That is, how many fetches and stores do you need to get the information into the places that are required to do the work? Again, if you look at polynomial evaluation, the tendency at present is to measure these things only by the number of arithmetic operations, but in an actual computer you have to do not only arithmetic operations, but you have to get the numbers to the right place to do the arithmetic. That, of course, is being ignored. Basically, the thing is, we're not equipped to deal with strong equivalence relations. We're dealing with almost the weakest equivalence relations we can and having trouble enough with that. Obviously, the distinction that a programmer makes in choosing between two algorithms for a particular problem is based on a very strong kind of equivalence relation. Typically, he will see distinctions between algorithms which are invisible from the viewpoint we are taking. Until we have developed theoretical tools that will analyze these things in terms of very strong equivalence relations, we can't really expect applications programmers to be very interested in our results.

KARP: I'd like to begin by suggesting that we need to find a name for our subject. "Computational complexity" is too broad in view of the work of Blum and others, at least until we can gather them into our fold; "Concrete computational complexity" is too much like civil engineering; "Complexity of computer computations" doesn't ring true; Markov has already taken "theory of algorithms" away from us; so I think it would be helpful in delineating our field for some clever guy to think of a name that we can all agree upon. In this connection, the matter of getting these materiels into university curricula, particularly the undergraduate curriculum, is important and I think it's going to happen very quickly.

As to the question of whether things are getting unified, I feel that there is considerable hope. We know how to go about things a little better now than we used to. It's very pleasing to see connections established with standard mathematics, theory of analytic functions, theory of quadratic forms, linear algebra, etc. As far as proof techniques and algorithm construction techniques go, I think we're learning. We now know that when we have a graph problem, we should immediately try depth first search before thinking any further. There are certain proof techniques pointed out by Strassen and Rabin, which sometimes allow us to show that there is no point in having to consider an adaptive algorithm, that straight-line nonadaptive algorithms will be as good as anything adaptive, and this is a very important conceptual aid in studying optimality questions. The notion of reducibility not only in the form that I used it, but in other connections, is important. For example, the results that Meyer and Fischer have observed showing that the transitive closure problem is of equivalent difficulty as the problem of multiplying Boolean matrices, gives us quite a different view than we might have had a priori of precisely where the difficulty resides in that problem. Other reducibilities of this type are very useful to look for in trying to see just where the hard part of a problem is. The so-called adversary approach in which we think of an algorithm as a dialogue between somebody who is executing the algorithm step by step by asking questions like, "Is this key bigger than that key or not?" and an adversary who tries to throw him off (a kind of a game theoretic approach to this worst case analysis of algorithms) is very important to keep in mind and very fruitful. The notion of introducing the kind of measure concerning how far along towards an answer you have progressed and then considering to what extent this measure can improve from step to step as exemplified by the togetherness measure that Floyd used in studying his file processing problem. This notion of attaching a numerical measure to the achievement that each step performs and hence finding the number of steps, I think is an important proof technique. The distinction between preconditioning and not preconditioning is now apparent to us, so I think we really know a lot more about how to go about these things.

Now, just one other brief remark, much of the work so far can be thought of in these terms -- what to do if you want to work in concrete computational complexity is you pick up a high school algebra book and look for an algorithm and then see if you can do it better than they do it in the book. That's quite useful and worthwhile, but maybe we shouldn't be looking just at high school algebra books. We should be looking at programming books. I think there are lots of things that computer people do that aren't mathematical in the nineteenth century sense. We manipulate strings, we prove theorems, we retrieve information, we manipulate algebraic formulas. I think by looking at the primitives appropriate to these domains, we can get a much richer collection of problems. The work of Morris and Pratt on a linear pattern matching algorithm is the kind of thing I have in mind. Some work that Hopcroft and I have done on efficient implementation of the unification algorithm in theorem proving is something that wouldn't have arisen as a question by looking at high school algebra, but does arise when you look at another domain of computation. I think when you do get into these other areas, more like programming than like ordinary mathematics, there are very severe definitional questions. One doesn't want to be premature in defining complexity because we might make the wrong definition, but on the other hand, when Mike Fischer makes the interesting conjecture, "Does there exist a linear time algorithm for the equivalence problem?" we're in an awkward position. We might be able to find one and thereby settle it, but if none does exist, we don't have a formal definition enabling us to prove that there is none. We don't have a good enough hold on the general concept of a data structure or an algorithm for this type of data manipulation. In that sense, I think some more methodological work needs to be done.

As for open problems, my favorite also is the question of whether deterministic polynomial bounded Turing machines are the same as nondeterministic ones.

FIDUCCIA: On the question of what we should call this area, let's take a cue from the titles of graduate courses being offered in this area. They seem to suggest something like "algorithm analysis and optimization." That is what I call my course, and I am suggesting that it is an appropriate name for this area.

On the question of whether a unified theory is developing from a scattering of lower bounds and algorithms, I feel optimistic. However, working against unification is the failure, in many cases, of specifying the model for the computation. As a result, the bounds obtained are often misunderstood. In order to have a well defined problem, one must specify the underlying mathematical system in which the computation is taking place, the operations which are allowed, the data that is given, and finally what is being computed. There seems to be too much emphasis on what is being computed and little emphasis on the model. Thus we speak of the complexity of a function as if it were an intrinsic property, when in fact we should be talking about the complexity of the task of computing the function from given data using the allowable operations in the particular underlying mathematical system. In specifying the model for computation, representation plays an important role. For example, a lower bound obtained by working at the word level does not apply if operations are allowed at the byte on bit level. In fact, having decided on a particular model for computation, it is best to completely leave the computer out of the analysis. If a unified theory is to develop, optimality results must be precisely stated -- nothing must be left to the imagination.

HOPCROFT: I would just like to comment a little bit about questions (1) and (2) and maybe expand on Floyd's answer to (1). Actually, I'm quite pleased at the way that theoretical results are developing and there is at least some sort of theory when you look at the development of algorithms, but if you look at how the theory of lower bounds is developing, I think you could change "slowly" to "it's not." As an example, Karp has shown that many of these problems that we've been looking at are complete, the implication being that therefore, these problems are hard. Well, that may not be the case. It may be that all of these problems are simple; in fact, there could even be algorithms which run in linear time, which solve all of these problems. In fact, given the present state of affairs, I think I would be reasonably safe in making the following conjecture; that is, that within maybe the next five years, nobody will prove that any of these problems takes more than let's say n^2 time. I think that's a reasonably safe conjecture and it also illustrates how little we know about lower bounds. On the other hand, the fact that we know so little about lower bounds, I think, provides a lot of interest in this area. I think that for many of the algorithms that have been improved recently, the improvements have come because the researchers became interested in the problems simply because they couldn't improve on the known lower bound. I know that this is true, for example, for the work that Tarjan and I did. The reason we considered connectivity at all was the thought that maybe there was something about connectivity of graphs which would help us to prove some sort of nontrivial lower bound on a computation. We didn't succeed, but we did find efficient algorithms. I think this is the reason for a large amount of interest in this area. Although Strassen can do matrix multiplication in 2.81 operations, nobody knows if it takes anything more than looking at each piece of data once to compute the product and I think this is what brings a lot of interest to the area.

MILLER: Thank you, gentlemen. I'm sure that each of you would like to ask each other some questions; however, that's unfair. You can talk among yourselves many times later, I'm sure. Let's use this opportunity to get some questions and comments from the audience.

RALPH WILLOUGHBY (Summary of question by MILLER): Essentially the question concerns stability and error analysis versus optimal algorithms, possibly not in a complexity sense.

RABIN: Well, from our experience, not too much was done about that. It is an extremely important issue, but we had the experience in some work that Sam [Winograd] and I did during the past year, that deeper understanding of complexity problems is a key also for treatment of this problem of numerical analysis. The methods for polynomial preconditioning and fast evaluation of polynomials were usually, if you look at some of the textbooks, discounted as leading to bad numerical results. Now, we worked on the problem of evaluating elementary functions, which eventually boils down to the question of finding the value of a rational function. We tried to use preconditioning to speed it up, and it turned out that if you just do it blindly, if you take any cookbook method for preconditioning, it really subverts your accuracy, but if you do it carefully, and you use all you know in an adaptive way, then you could, in fact, combine considerable time savings with the original accuracy, which in the example which we worked out was last bit accuracy in double precision evaluation. Now, over the whole range, I think that is a big problem. However, the problem of accuracy wouldn't come up, of course, if you go to the non-numerical applications, which I was advocating before, so that may be another incentive for going into non-numerical problems.

Since I have the microphone, I would like to give an answer to Dick Karp's question. If we are going to expropriate "computational complexity" from the recursive function people, why not go all the way and expropriate "computability" as a name for our field? I'm not saying it completely seriously, but it's a possibility.

DAVID SAYRE (Summary of question by WINOGRAD): If I understood the question correctly, the question was, "Are there any results about measuring not the time or the number of iterations, but the length of a program needed to solve that problem?"

I'm going to answer that in very simple terms -- no, not as far as I know. There are many questions around that I could ask. One could point out that there is very little known about the length of a program, very little known about the bookkeeping part of the program, very little known about size of storage which is needed. If you start looking around, I think that the amount which is unknown by far surpasses the amount of what we know. I think that one of the panelists said before that one of the important problems here is to find a very precise mathematical formulation of the problem. Maybe that is the reason that arithmetic operations, which we knew how to tie more immediately to known mathematics, were attacked first. I think that we are right now at this point where people will try to formulate and make precise questions relating to other areas. So to my knowledge, the answer to the question is, right now no, and I want to add that I hope at the next conference we'll be able to say yes.

HOPCROFT: I'd like to say one thing about that problem. If you take a complex problem and try to find the optimal algorithm, I suspect that there is a very high probability that this optimal algorithm will be much shorter, the actual number of statements, etc., than the algorithm that you might at first-hand use. The reason is that usually an optimal algorithm has some nice structure to it and this simplifies the programming tremendously. One thing about algorithms, once you see how to do the problem right, quite often it is easier to show that the optimal algorithm is correct rather than the first-hand crack that you might take.

KARP: If I can add just one further comment, I'm not convinced of the notion that the length of a program is a measure at all of the difficulty of writing it. It seems to me it's much easier to write a long program to do something than to write a short one.

DONALD ROSE: There are two differences between the Sparse Matrix Symposium and this one, but the major difference is the engineers at the Sparse Matrix Symposium were really in there and here the engineers seemed to be hiding. That's significant, I think, with respect to the question which was asked, "What does all this mean as far as actual computation?" Secondly, the interaction between the audience at this Symposium and the speakers seems far less robust than at the Sparse Matrix Symposium. That's unfortunate, too.

With respect to this question about, "Will these computational complexity results actually produce important applied results?" let's say that, I think that the answer is assuredly yes and, in fact, I will give an example of that in a minute. However, I agree with the panel that probably the most important question to answer is the completeness question. I must admit that at this conference was the first time that I had ever heard of that question. I think it's fantastically important.

I want to say that there is a result, I think a significant practical result, that may come out of this meeting. I spent a considerable amount of time talking with Robert Tarjan, and the fact that you can now compute the strong components of a directed graph in a time linear in the number of edges, or linear in the number of vertices for the types of graphs that we would consider, is a very impressive result. This means that input/output systems (of order thousand by thousand, say), which have certain structure, may be shown to be solvable in linear time, and since such problems are going to be solved more and more, I think that's significant. The point is that in present-day numerical analysis, if you're talking about polynomial time, it better be polynomial of degree less than five.

Now, what do I mean by the problems that arise? That's the crucial question and the point is if, in fact, it turns out that there are "hard" problems (i. e., not all hard problems can be shown to be "easy"), then this question becomes part of a more pressing problem: how do we find the subset of these hard problems which we can solve in efficient time? This more general consideration tries to get at the question, "Will it all be useful?" And I think the answer is definitely yes. But there are times when we're just going to have to really restrict the class of problems that we're working on.

MILLER: Thank you, Don. Maybe the panel has some reply to that. Does it need a reply?

ALBERT MEYER: We might as well continue with this robustness. I'd like to disagree mildly with a couple of things that the final panelists said about machine models, essentially reiterating John's [Hopcroft] point, with which I agree. The issue of finding a realistic machine model is one we have to face. There has been some study of formulations of random access machines and their computational properties, but the fact is that that's not the real issue in getting lower bounds. In order to prove that a certain problem is hard, in the sense that it takes a lot of time, the obstacle is not that we don't have a good formulation of a machine model (as it was, for example, in computability theory where one defends Church's thesis by evoking a large number of definitions of "algorithm" and convincing oneself that they are all the same; hence, "algorithms" are a well defined concept, and then one can proceed to prove negative results about this well defined mathematical class). We have many different models, but we can't prove lower bounds for any of them, and so the issue is not finding the right model in order to establish our lower bound results in general and make them really persuasive. Take a perfectly well defined mathematical model like a multitape Turing machine, which I do not defend as being realistic, but I do defend as being simple and well defined. I will predict that any kind of model that you claim is realistic will be more complicated than a multitape Turing machine. I'm offering this multitape Turing machine as a model that we now have, even though we're not satisfied with its generality. We cannot prove any interesting lower bounds about this machine. There is no known recognition problem which we can prove requires more than twice the time required to read the input. That's the state of our ignorance. This is a sobering note, which should be kept in mind, particularly in view of the emphasis on the problem of whether $P = NP$ that is, whether polynomial time and nondeterministic polynomial time are different. We can't even begin to deal with polynomials if we can't deal with time $2n$.

KARP: I agree with what Albert [Meyer] says in part; however, it's important to be aware that for many specific problems, a reasonable ad hoc model can be set up for which it is possible to prove lower bounds. We have many examples: the addition and multiplication chains that were studied by Winograd and others in connection with polynomial evaluation; the sequences of comparisons in connection with sorting; the primitive rearrangement steps in Floyd's algorithms, the ones he discussed this morning. These examples can be multiplied. It's not entirely satisfying, of course, to work with a restricted class of algorithms rather than the entire class representable, say, on a random access machine, but nevertheless, by suitable, reasonable restrictions, one can very often get interesting lower bounds. So, yes, there is a problem with lower bounds, but it's not entirely hopeless if we're willing to make some asumptions at times.

WINOGRAD: Let me agree completely with Dick [Karp] and

emphasize it even more. I believe that maybe we don't know much about two-tape Turing machines. Maybe as long as we are civil engineers and talking about concrete complexity, maybe we are not even interested in the generality that two-tape Turing machines represent; maybe on the other hand, if we understand more about actual sorting, not just the comparisons, but also how we move data from one place to another. That is, set up a specific problem and formulate it (and here I include storage and the movement of the data) and then say something meaningful about that specific problem. I think this is more the name of the game called concrete complexity. The problems we solve will be smaller, will not be on a grand scale, but maybe that's the reason I'm optimistic that we will be able to solve it while the two-tape Turing machine, with its enormity and a lot of things that can happen there, will not be solved.

STRASSEN: I'd just like to state a result which has recently been obtained by Stoss, a student of Schonhage. It concerns the problem of permuting a sequence of n free symbols on a multitape Turing machine. So the machine may print such a symbol on a tape square only if at the same time it sees the same symbol on some other square. Stoss derives lower bounds of the order $n \log n$ for specific permutations, even if the control of the machine is infinite. I think that's a very encouraging step in the direction of getting nontrivial lower bounds for the computation time on multitape Turing machines.

RABIN: I think that we have really two issues here, which sort of get intertwined. One is the question of specific algorithms and there, the rules of the game are very important and you have to devise a model for the given situation. If, for example, you are studying sorting, internal sort is different than what you have when you consider sort from tapes or maybe from a disc. That's one of the troublesome aspects of this area, that it is quite model-dependent within certain limits. The question whether something is of complexity n, or n^2, or $n \log n$, or even whether it is $3n$ or $5n$ is extremely important. You can't decide such questions without having a rather specific model. Even so, coming back to the remark I made at the beginning, I'm hopeful about the fact there are certain general observations and methods. We forgot to mention the idea of preconditioning. This is well known from polynomials -- that you do something with the coefficients before you do many repeated calculations. But if you examine other algorithms, even if they are done just once, you may very often discover that what is actually happening is that small parts of the data are being preconditioned for later use within this algorithm, so that is an example of the general methodology and philosophy which is useful in many instances. Coming back to the $P = NP$ problem, there it is true that we are not dependent on the model. I also agree that we are hard pressed to find something which really grows faster in terms of complexity than the actual data. However, I'm not without hope. I think we need some additional ideas. I think that the main problem perhaps is that most of the methods, with one or two exceptions, which were used in the past in order to prove that one problem is more difficult than another problem, actually used diagonalization. That was really the main method which was borrowed from the theory of recursive functions. It is strong enough in some situations, but may not be strong enough in this particular situation. It is a hard problem, it is an unusual problem, and it will require unusual intellect and perseverence to do it. I can't really say whether I'm with John [Hopcroft], who, I think, has put his money on the possibility of polynomial complexity of all those problems, which, if true, would be sensational, or with the other camp. Finally, it is also possible that these problems will find a resolution and a solution through non-machine methods, through methods which involve analysis and somewhat more classical mathematics. The last mentioned methods are powerful, and it is a question of tying the two things together.

MILLER: In virtue of time, I'm going to let Mike Paterson have the mike for a few minutes. He's been champing at the bit, I see.

PATERSON: I've stopped champing at the bit now, because I think Michael [Rabin] has been covering it fairly well. Lots of people have hinted at and talked around the short point I want to make, but I haven't heard anybody come out and say it really bluntly and boldly. That point is that it's the solution which is of prime importance rather than the problem and that we should tackle the solutions first and then wrap problems around them. The very best of problems, if it happens to fall to a boring technique, was the wrong problem. Suppose somebody were to prove $P \neq NP$ then this would be of great personal and dramatic interest, in view of the people that have tried to work on it. But if it were to be established by a quite traditional, ordinary argument, which it just happened that nobody had thought of, then we would see in retrospect that it was the wrong problem. We might not be particularly interested in fast algorithms for multiplying very large integers or fast algorithms for multiplying matrices, unless the techniques used have independent interest. So I'd like to declare for the solution rather than the problem.

MILLER: Thank you, Mike [Paterson]. In view of the time and knowing that a lot of people have to get out of here, I'm going to close the session at this time. I would like to say, however, that I would like to harken back to Don Rose's comment about interaction with engineers. I would say "practitioners" possibly would be a better term. I think there's a tremendous amount of promise here; that is, that interaction could be very fruitful, and I'm sure that the people working in this area are eagerly looking for this. Thank you.

BIBLIOGRAPHY

Aanderaa, S. O.
(1969A) See Cook and Aanderaa.

Arbib, M. A.
(1967A) See Spira and Arbib.

(1969A) Theories of Abstract Automata, Prentice-Hall, Englewood Cliffs, New Jersey.

Ariyoshi, A., Shirakawa, I., and Hiroshi, O.
(1971A) "Decomposition of a graph into compactly connected two-terminal subgraphs," IEEE Transactions on Circuit Theory, Vol. CT-18, No. 4, 430-435.

Arlazarov, V. L., Dinic, E. A., Kronrod, M. A., and Faradzhev, I. A.
(1970A) "On economical construction of the transitive closure of an oriented graph," Dokl. Akad. Nauk SSSR, Vol. 194, 487-488 (Russian). English translation in Soviet Math. Dokl., Vol. 11, 1209-1210.

Barnes, J. P. G.
(1965A) "An algorithm for solving non-linear equations based on the secant method," Comp. J., Vol. 8, 66-72.

Batcher, K. E.

(1968A) "Sorting networks and their applications," Proceedings AFIPS Spring Joint Computer Conference, Vol. 32, 307-314.

Belaga, E. G.

(1958A) "Some problems in the computation of polynomials," Dokl. Akad. Nauk SSSR, Vol. 123, 775-777 (Russian). See Math. Reviews, Vol. 21 (1960) review number 3935.

(1961A) "On computing polynomials in one variable with initial conditioning of the coefficients," Problemy Kibernet., Vol. 5, 7-15 (Russian). English translation in Problems of Cybernetics, Vol. 5, 1-13.

Bittner, L.

(1959A) "Eine Verallgemeinerung des Lekantenverfahrens (regula falsi) zur naherungsweisen Breechnung der Nullstellen eines nichtlinearen Gleichungssystems," Wissen. Zeit. der Technischen Hochschule Dresden, Vol. 9, 325-329.

Blevins, P. R.

(1971A) See Ramamoorthy and Blevins.

Borodin, A.

(1971A) "Horner's rule is uniquely optimal," Theory of Machines and Computations, edited by Z. Kohavi and A. Paz, Academic Press, New York, 45-58.

(1972A) "Computational complexity -- theory and practice," to appear in Current Trends in the Theory of Computing, edited by A. Aho, Prentice-Hall, Englewood Cliffs, New Jersey.

Borodin, A., and Munro, I.

(1971A) "Evaluating polynomials at many points," Information Processing Letters, Vol. 1, 66-68.

Brauer, A.
(1939A) "On addition chains," Bull. Amer. Math. Soc., Vol. 45, 736-739.

Brent, R. P.
(1970A) "Error analysis of algorithms for matrix multiplication and triangular decomposition using Winograd's identity," Numer. Math., Vol. 16, 145-156.

(1970B) "Algorithms for matrix multiplication," Technical Report STAN-CS-70-157, Department of Computer Science, Stanford University, Stanford, California.

(1972A) Algorithms for Minimization without Derivatives, Prentice-Hall, Englewood Cliffs, New Jersey (Chapter 3).

(1972B) "On maximizing the efficiency of algorithms for solving systems of nonlinear equations," IBM Watson Research Center, Yorktown Heights, New York, RC 3725.

Brown, K. M., and Conte, S. D.
(1967A) "The solution of simultaneous nonlinear equations," Proceedings 22nd National Conference of the ACM, Thompson Book Co., Washington, D. C., 111-114.

Busacker, R. G., and Saaty, T. L.
(1965A) Finite Graphs and Networks: An Introduction with Applications, McGraw Hill, New York, 196-199.

Cohen, A. I., and Varaiya, P.
(1970A) "Rate of convergence and optimality conditions of root finding," to appear.

Conte, S. D.
(1967A) See Brown and Conte.

Cook, S. A.
(1966A) "On the minimum computation time of functions," Doctoral Thesis, Harvard University, Cambridge, Massachusetts.

(1970A) "The complexity of theorem-proving procedures," Proceedings Third Annual ACM Symposium on the Theory of Computing, Shaker Heights, Ohio, 151-158.

Cook, S. A., and Aanderaa, S. O.
(1969A) "On the minimum computation time of functions," Trans. Amer. Math. Soc., Vol. 142, 219-314.

Corneil, D. G.
(1971A) "An n^2 algorithm for determining the bridges of a graph," Information Processing Letters, Vol. 1, 51-55.

Davenport, H.
(1970A) Higher Arithmetic: The Introduction to the Theory of Numbers, Hutchinson, London.

deBoor, C., and Fix, G.
(1972A) "Spline approximation by quasi-interpolants," J. Approx. Theory, to appear.

Dijkstra, E.
(1959A) "A note on two problems in connection with graphs," Num. Math., Vol. 1, 269-271.

Dinic, E. A.
(1970A) See Arlazarov, Dinic, Kronrod, and Faradzhev.

Dorn, W. S.
(1962A) "Generalizations of Horner's rule for polynomial evaluation," IBM J. of Research and Development, 239-245.

Dorr, F.
(1970A) "The direct solution of the discrete Poisson equation on a rectangle," SIAM Review, Vol. 12, 248-263.

Edmonds, J.
(1965A) "Paths, trees and flowers," Canadian J. Math. XVII, 449-467.

Edmonds, J., and Karp, R.
(1972A) "Theoretical improvements in algorithmic efficiency for network flow problems," J ACM.

Eisenstat, S. C., and Schultz, M. H.
(1972A) "Computational aspects of the finite element method," Proceedings of the Symposium on the Mathematical Foundations of the finite element method with application to partial differential equations, to appear.

Emel'yanov, K. V., and Ill'in, A. M.
(1967A) "Number of arithmetical operations necessary for the approximate solution of Fredholm integral equations of the second kind," Zh. Vychisl. Mat. i Mat. Fiz., Vol. 7, 905-910 (Russian). English translation in USSR Comput. Math. and Math. Phys., Vol. 7, 259-266.

Eve, J.
(1964A) "The evaluation of polynomials," Numer. Math., Vol. 6, 17-21.

Faradzhev, I. A.
(1970A) See Arlazarov, Dinic, Kronrod, and Faradzhev.

Feldstein, A., and Firestone, R. M.
(1969A) "A study of Ostrowski efficiency for composite iteration functions," Proceedings ACM National Conference, 147-155.

Fiduccia, C. M.
(1971A) "Fast matrix multiplication," Proceedings of Third Annual ACM Symposium on Theory of Computing, 45-49.

Fike, C. T.
(1969A) See Sterbenz and Fike.

Firestone, R. M.
(1969A) See Feldstein and Firestone.

Fischer, M. J.
(1964A) See Galler and Fischer.

Fischer, M. J., and Meyer, A. R.
(1971A) "Boolean matrix multiplication and transitive closure," IEEE Conference Record of the Twelfth Annual Symposium on Switching and Automata Theory, 129-131.

Fix, G.
(1968A) "Higher-order Rayleigh-Ritz approximations," J. Math. Mech., Vol. 18, 645-658.

(1972A) See deBoor and Fix.

Fix, G., Gulati, S., and Wakoff, G.
(1972A) "On the use of singular functions with finite element approximations," to appear.

Floyd, R. W., and Knuth, D. E.
(1970A) "The Bose-Nelson sorting problem," CS Report 70-177, Stanford University, Stanford, California.

Ford, L. R., Jr., and Johnson, S. M.
(1959A) "A tournament problem," Amer. Math. Monthly, Vol. 66, 387-389.

Furman, M. E.
(1970A) "Application of a method of fast mutliplication of matrices in the problem of finding the transitive closure of a graph," Dokl. Akad. Nauk. SSSR, Vol. 194, 524 (Russian). English translation in Soviet Math. Dokl., Vol. 11, 1252.

Galler, B. A., and Fischer, M. J.
(1964A) "An improved equivalence algorithm," Comm. ACM, Vol. 7, No. 5, 301-303.

Gentleman, M. W.
(1971A) Private communication.

George, J. A.

(1971A) "Computer implementation of the finite element method," Doctoral Thesis, Stanford University, Stanford, California.

(1972A) "Block elimination on finite element systems of equations," Proceedings of the IBM Symposium on Sparse Matrices and Their Applications, Plenum Press, New York, New York.

Graham, R. L.

(1971A) "Sorting by comparisons," Computers in Number Theory, edited by A. O. L. Atkin and B. J. Birch, Academic Press, New York, 263-269.

Green, M. W.

(1970A) "Some observations on sorting," Cellular Logic in Memory Arrays, Final Report, Part 1, SRI Project 5509, Stanford Research Institute, Menlo Park, California, 49-71.

Grossman, D. D., and Silverman, H. F.

(1971A) "Placement of records on a secondary storage device to minimize access time," IBM Watson Research Center, Yorktown Heights, New York, RC 3641.

Gulati, S.

(1972A) See Fix, Gulati, and Wakoff.

Hadian, A.

(1969A) "Optimality properties of various procedures for ranking n different numbers using only binary comparisons," Technical Report No. 117, Department of Statistics, University of Minnesota, Minneapolis, Minnesota.

Hadian, A., and Sobel, M.

(1969A) "Selecting the tth largest using binary errorless comparisons," Colloquia Mathematica Societatis Janos Bolyai, Balatonfured, Hungary, 585-599.

Hadian, A., and Sobel, M.
(1970A) "Ordering the t largest of n items using binary comparisons," Proceedings Second Chapel Hill Conference on Combinatorial Mathematics and Its Applications.

Hall, M.
(1948A) "Distinct representatives of subsets," Bull. Amer. Math. Soc., Vol. 54, 922-926.

Harary, F.
(1969A) Graph Theory, Addison-Wesley, Reading, Massachusetts.

Hardy, G. H., Littlewood, J. E., and Polya, G.
(1926A) "The maximum of a certain bilinear form," Proceedings London Math. Soc. (2), Vol. 25 265-282.

(1934A) Inequalities, Cambridge University Press, Cambridge, England.

Hardy, G. H., and Wright, E. M.
(1938A) An Introduction to the Theory of Numbers, Clarendon Press, Oxford, Thm. 354.

(1960A) The Theory of Numbers, Oxford, Chapter XI.

Harper, L. H.
(1970A) "Combinatorial coding theory," Proceedings Second Chapel Hill Conference on Combinatorial Mathematics and Its Applications, University of North Carolina, 252-260.

Hiroshi, O.
(1971A) See Ariyoshi, Shirakawa, and Hiroshi.

Hoare, C. A. R.
(1962A) "Quicksort," Comput. J., Vol. 5, 10-15.

Holt, R. C., and Reingold, E. M.
(1970A) "On the time required to detect cycles and connectivity in directed graphs," Technical

Holt and Reingold (cont'd)
Report No. 7-63, Department of Computer Science, Cornell University, Ithaca, New York, to appear in Math. Systems Theory (1972).

Hopcroft, J.
(1971A) "An N log N algorithm for isomorphism of planar triply connected graphs," Technical Report STAN-CS-71-192, Computer Science Department, Stanford University, Stanford, California.

(1971B) "An N log N algorithm for minimizing states in a finite automaton," Theory of Machines and Computations, Academic Press, New York, 189-196.

Hopcroft, J., and Karp, R. M.
(1971A) "A $n^{5/2}$ algorithm for maximum matchings in bipartite graphs," IEEE Conference Record of the Twelfth Annual Symposium on Switching and Automata Theory, 122-125.

Hopcroft, J., and Kerr, L.
(1969A) "Some techniques for proving certain simple programs optimal," IEEE Conference Record of the Tenth Annual Symposium on Switching and Automata Theory, 36-45.

(1971A) "On minimizing the number of multiplications necessary for matrix multiplication," SIAM J. Appl. Math., Vol. 20, 30-36.

Hopcroft, J., and Tarjan, R.
(1971A) "A V^2 algorithm for determining isomorphism of planar graphs," Information Processing Letters, Vol. 1, North Holland Publishing Company, Amsterdam, 32-34.

(1971B) "Efficient algorithms for graph manipulation," Technical Report STAN-CS-71-207, Computer Science Department, Stanford University (to appear in C ACM).

Hopcroft and Tarjan (cont'd)

(1971C) "Planarity testing in V log V steps," Information Processing 1971, Proceedings of IFIP Congress 71.

(1972A) "An O(V) planarity algorithm," to be published.

Hopcroft, J., and Ullman, J. D.

(1969A) Formal Languages and Their Relation to Automata, Addison-Wesley, Reading, Massachusetts.

(1971A) "A linear list merging algorithm," Technical Report TR 71-111, Computer Science Department, Cornell University, Ithaca, New York.

Hwang, F. K., and Lin, S.

(1969A) "An analysis of Ford and Johnson's sorting algorithm," Proceedings of the Third Annual Princeton Conference on Information Sciences and Systems, 292-296.

(1971A) "Optimal merging of 2 elements with n elements," Acta Informatica, Vol. 1, 145-158.

Il'in, A. M.

(1967A) See Emel'yanov and Il'in.

Johnson, S. M.

(1959A) See Ford and Johnson.

Karatsuba, A., and Ofman, Yu

(1962A) "Multiplication of multidigit numbers on automata," Dokl. Akad. Nauk SSSR, Vol. 145, 293-294 (Russian). English translation in Soviet Physics Dokl., Vol. 7 (1963) 595-596.

Karp, R. M.

(1971A) See Hopcroft and Karp.

(1971B) Notes from CS 230, Berkeley, California, unpublished.

Karp (cont'd)

(1972A) See Edmonds and Karp.

(1972B) "Reducibility among combinatorial problems," this conference.

Karp, R. M., and Miranker, W. L.

(1968A) "Parallel minimax search for a maximum," J. of Combinatorial Theory, Vol. 4, 19-35.

Kerr, L. R.

(1969A) See Hopcroft and Kerr.

(1969B) See Hopcroft and Kerr.

(1970A) "The effect of algebraic structure on the computational complexity of matrix multiplication," Doctoral Thesis, Cornell University, Ithaca, New York.

(1971A) See Hopcroft and Kerr.

Kiefer, J.

(1953A) "Sequential minimax search for a maximum," Proceedings Amer. Math. Soc., Vol. 4, 502-506.

(1957A) "Optimum sequential search and approximation methods under minimum regularity assumptions," J. Soc. Indust. Appl. Math., Vol. 5, 105-137.

King, R. F., and Phillips, D. L.

(1969A) "The logarithmic error and Newton's method for the square root," Comm. ACM, Vol. 12, 87-88.

Kislicyn, S. S.

(1962A) "On a bound for the least mean number of pairwise comparisons needed for a complete ordering of n objects with different weights," Vestnik Leningrad, Univ. Ser. Mat. Meh. Astronom., Vol. 17, 162-163 (Russian with English summary).

Kislicyn (cont'd)

(1964A) "On the selection of the kth element of an ordered set by pairwise comparisons," Sibirsk. Mat. Zh., Vol. 5, 557-564 (Russian). See Math. Reviews, Vol. 29, review number 2198.

Kljuev, V. V., and Kokovkin-Shcherbak, N. I.

(1965A) "On the minimization of the number of arithmetic operations for the solution of linear algebraic systems of equations," Zh. Vychisl. Mat. i Mat. Fiz., Vol. 5, 21-23 (Russian). English translation in USSR Comput. Math. and Math. Phys., Vol. 5, 25-43. Also translated in Technical Report CS24, Department of Computer Science, Stanford University, Stanford, California.

(1966A) "Minimization of the number of arithmetical operations in a transformation of matrices," Ukrain. Mat. Zh., Vol. 18, 122-128 (Russian). See Math Reviews, Vol. 34, review number 3767.

(1967A) "Minimization of computational algorithms in certain transformations of matrices," Zh. Vychisl. Mat. i Mat. Fiz., Vol. 7, 3-13 (Russian). English translation in USSR Comput. Math. and Math. Phys., Vol. 7, 1-14.

Knuth, D. E.

(1962A) "Evaluation of polynomials by computers," Comm. ACM., Vol. 6, 595-599.

(1968A) The Art of Computer Programming, Vol. 1, Addison-Wesley, Reading, Massachusetts.

(1969A) The Art of Computer Programming, Vol. 2, Addison-Wesley, Reading, Massachusetts.

(1970A) See Floyd and Knuth.

Knuth (cont'd)

(1971A) "Mathematical analysis of algorithms," Technical Report No. STAN-CS-71-206, Department of Computer Science, Stanford University, Stanford, California.

(1972A) "Some combinatorial research problems with a computer-science flavor," notes by L. Guibas and D. Plaisted from an informal seminar, January 17, 1972.

(1972B) The Art of Computer Programming, Vol. 3, Addison-Wesley, Reading, Massachusetts.

Kokovkin-Shcherbak, N. I.

(1965A) See Kljuev and Kokovkin-Shcherbak.

(1966A) See Kljuev and Kokovkin-Shcherbak.

(1967A) See Kljuev and Kokovkin-Shcherbak.

(1968A) "Minimization of computational algorithms in the solution of the elimination problem," Zh. Vychisl. Mat. i Mat. Fiz., Vol. 8, 1096-1101 (Russian). English translation in USSR Comput. Math. and Math. Phys., Vol. 8, 212-219.

(1970A) "On minimization of calculation algorithms in solutions of arbitrary systems of linear equations," Ukrain. Mat. Zh., Vol. 22, 494-502 (Russian).

Kronrod, M. A.

(1970A) See Arlazarov, Dinic, Kronrod, and Faradzhev.

Kruskal, J.

(1956A) "On the shortest spanning subtree of a graph and the traveling salesman problem," Proceedings Amer. Math. Soc., Vol. 7, 48-50.

Lawler, E. L.

(1971A) "The complexity of combinational computations: a survey," Proceedings of 1971 Polytechnic Institute of Brooklyn Symposium on Computers and Automata.

Lederberg, J.

(1964A) "Dendral-64: A system for computer construction, enumeration, and notation of organic molecules as tree structures and cyclic graphs, Part I: Notational algorithm for tree structures," Interim Report to the National Aeronautics and Space Administration, Grant NSG 81-60, NASA Cr 68898, STAR No. N-66-14074.

Lehman, R. S.

(1959A) "Developments at an analytic corner of solutions of elliptic partial differential equations," J. Math. Mech., Vol. 8, 727-760.

Lin, S.

(1969A) See Hwang and Lin.

(1971A) See Hwang and Lin.

Littlewood, J. E.

(1926A) See Hardy, Littlewood, and Polya.

(1934A) See Hardy, Littlewood, and Polya.

Liu, C. L.

(1971A) "Analysis of sorting algorithms," IEEE Conference Record of the Twelfth Annual Symposium on Switching and Automata Theory, 207-215.

Maruyama, K.

(1971A) "Parallel methods and bounds of evaluating polynomials," Report No. 437, Department of Computer Science, University of Illinois, Urbana, Illinois.

Meyer, A. R.
(1971A) See Fischer and Meyer.

Miranker, W. L.
(1968A) See Karp and Miranker.

Moon, J. W.
(1968A) Topics on Tournaments, Holt, Rinehart, and Winston, New York.

Morgenstern, J.
(1971A) "On linear algorithms," Theory of Machines and Computations, edited by Z. Kohavi and A. Paz, Academic Press, New York, 59-66.

Morris, R.
(1969A) "Some theorems on sorting," SIAM J. Appl. Math., Vol. 17, 1-6.

Motzkin, T. S.
(1955A) "Evaluation of polynomials" and "Evaluation of rational functions," Bull. Amer. Math. Soc., Vol. 61, 163 (abstracts).

Moursund, D. G.
(1967A) "Optimal starting values for Newton-Raphson calculation of $x^{1/2}$," Comm. ACM, Vol. 10, 430-432.

Munro, I.
(1971A) See Borodin and Munro.

(1971B) "Some results concerning efficient and optimal algorithms," Proceedings of Third Annual ACM Symposium on Theory of Computing, 40-44.

(1971C) "Some results in the study of algorithms," Technical Report No. 32, Department of Computer Science, University of Toronto, Toronto, Ontario, Canada.

Munro (cont'd)
(1971D) "Efficient determination of the transitive closure of a directed graph," Information Processing Letters, Vol. 1, 56-58.

Munro, I., and Paterson, M.
(1971A) "Optimal algorithms for parallel polynomial evaluation," Proceedings Twelfth Annual Symposium on Switching and Automata Theory, 132-139.

Munteanu, J. M., and Schumaker, L.
(1972A) Private communication.

Muraoka, Y.
(1971A) "Parallelism exposure and exploitation in programs," Report No. 424, Department of Computer Science, University of Illinois, Urbana, Illinois.

Ofman, Yu
(1962A) See Karatsuba and Ofman.

(1962B) "On the algorithmic complexity of discrete functions," Dokl. Akad. Nauk SSSR, Vol. 145, 48-51 (Russian). English translation in Soviet Physics Dokl., Vol. 7 (1963) 589-591.

Ostrowski, A. M.
(1954A) "On two problems in abstract algebra connected with Horner's rule," Studies in Mathematics and Mechanics, Presented to R. von Mises, Academic Press, New York, 40-48.

(1966A) Solution of Equations and Systems of Equations, Academic Press, New York.

Pan, V. Ya
(1959A) "Schemes for computing polynomials with real coefficients," Dokl. Akad. Nauk SSSR, Vol. 127, 266-269 (Russian). See Math. Reviews, Vol. 23 (1962), review number B560. French translation available from Laboratoire Central

Pan (cont'd)

De L'Armement Traductions, 16 bix Avenue Arcueit, France; translation number T-1.524.

(1961A) "Certain schemes for the evaluation of polynomials with real coefficients," Problemy Kibernet., Vol. 5, 17-29 (Russian). English translation in Problems of Cybernetics, Vol. 5 (1964) 14-32.

(1962A) "On several ways of computing values of polynomials," Problemy Kibernet., Vol. 7, 21-30 (Russian). See Comput. Rev., Vol. 3, review number 3329.

(1962B) "Schemes using preliminary treatment of coefficients for polynomial calculation. A program for automatic determination of parameters," Zh. Vychisl. Mat. i Mat. Fiz., Vol. 2, 133-140 (Russian). English translation in USSR Comput. Math. and Math. Phys., Vol. 2, 137-146.

(1965A) "The computation of polynomials of fifth and seventh degree with real coefficients," Zh. Vychisl. Mat. i Mat. Fiz., Vol. 5, 116-118 (Russian). English translation in USSR Comput. Math. and Math. Phys., Vol. 5, 159-161.

(1966A) "Methods of computing values of polynomials," Uspehi. Mat. Nauk, Vol. 21, 103-134 (Russian). English translation in Russian Math. Surveys, Vol. 21, 105-136.

Paterson, M.

(1971A) See Munro and Paterson.

(1971B) "Notes on optimality for square root algorithms," unpublished.

Paterson, M., and Stockmeyer, L.
(1971A) "Bounds on the evaluation time for rational polynomials," IEEE Conference Record of the Twelfth Annual Symposium on Switching and Automata Theory, 140-143.

Phillips, D. L.
(1969A) See King and Phillips.

Pohl, I.
(1969A) "A minimum storage algorithm for computing the median," Report No. RC 2701, IBM Watson Research Center, Yorktown Heights, New York.

Polya, G.
(1926A) See Hardy, Littlewood, and Polya.

(1934A) See Hardy, Littlewood, and Polya.

Rabin, M. O.
(1971A) "Proving simultaneous positivity of linear forms," Third Annual ACM Symposium on Theory of Computing, invited address.

(1972A) "Solving a system of linear equations with unknown coefficients," this conference.

Ramamoorthy, C. F., and Blevins, P. R.
(1971A) "Arranging frequency dependent data on sequential memories," Proceedings AFIPS SJCC, Vol. 38, 545-556.

Reingold, E. M.
(1970A) See Holt and Reingold.

(1970B) "On the optimality of some set and vector algorithms," IEEE Conference Record of the Eleventh Annual Symposium on Switching and Automata Theory, 68-71.

Reingold (cont'd)

(1971A) "On some optimal algorithms," Doctoral Thesis, Cornell University, Ithaca, New York. Also available as Report No. 428 of the Department of Computer Science, University of Illinois, Urbana, Illinois.

(1971B) "Computing the maximum and the median," IEEE Conference Record of the Twelfth Annual Aymposium on Switching and Automata Theory, 216-218.

Rissanen, J.

(1971A) "On optimum root-finding algorithms," J. Math. Anal. Appl., Vol. 36, 220-225.

Rogers, H.

(1967A) Theory of Recursive Functions and Effective Computability, McGraw-Hill, New York.

Saaty, T. L.

(1965A) See Busacker and Saaty.

Salomaa, A.

(1969A) Theory of Automata, Pergamon Press, Elmsford, New York.

Sandelius, M.

(1961A) "On an optimal search procedure," Amer. Math. Monthly, Vol. 68, 133-134.

Schonhage, A.

(1966A) "Multiplication of large numbers," Computing, Vol. 1, 182-196 (German with English summary). See Comput. Rev., Vol. 8 (1967) review number 11544.

(1971A) "Fast computation of continued fraction expansions," Acta Informatica, Vol. 1, 139-144 (German).

Schonhage, A., and Strassen, V.
(1971A) "Fast multiplication of large numbers," Computing, Vol. 7, 281-292 (German with English summary).

Schreier, J.
(1932A) "On tournament eliminations systems," Mathesis Polska, Vol. 7, 154-160 (Polish).

Schultz, M. H.
(1971A) "L^2-error bounds for the Rayleigh-Ritz-Galerkin method," SIAM J. Num. Anal., Vol. 8 737-748.

(1972A) See Eisenstat and Schultz.

(1972B) "Quadrature-Galerkin approximations to solutions of elliptic differential equations," Proceedings of the Amer. Math. Soc., to appear.

(1973A) Finite Element Analysis, Prentice-Hall, Inc., Englewood Cliffs, New Jersey, to appear.

Schumaker, L.
(1972A) See Munteanu and Schumaker.

Scoins, H. I.
(1968A) "Placing trees in lexicographic order," Machine Intelligence 3, American Elsevier Publishing Co., New York, 43-60.

Shirakawa, I.
(1971A) See Ariyoshi, Shirakawa, and Hiroshi.

Silverman, H. F.
(1971A) See Grossman and Silverman.

Slupecki, J.
(1951A) "On the system S of tournaments," Colloq. Math., Vol. 2, 286-290.

Smith, C. A. B.

(1947A) "The counterfeit coin problem," Math. Gaz., Vol. 31, 31-39.

Sobel, M.

(1968A) "On an optimal search for the t best using only binary errorless comparisons: the ordering problem," Technical Report No. 113, Department of Statistics, University of Minnesota, Minneapolis, Minnesota.

(1968B) "On an optimal search for the t best using only binary errorless comparisons: the selection problem," Technical Report No. 114, Department of Statistics, University of Minnesota, Minneapolis, Minnesota.

(1969A) See Hadian and Sobel.

(1970A) See Hadian and Sobel.

Spira, P. M.

(1968A) "On the computation time of finite functions," IEEE Conference Record of the Ninth Annual Symposium on Switching and Automata Theory, 69-75.

(1968B) "On the time required for group multiplication," Proceedings of the Hawaii International Conference in Systems Science, 316-319.

(1969A) "The time required for group multiplication," J. Assoc. Comput. Mach., Vol. 16, 235-243.

(1969B) "On the computation time of certain classes of Boolean functions," Conference Record of ACM Symposium on Theory of Computing, 271-272.

(1969C) "On the complexity of group multiplication with constant coding," Proceedings of the Third Annual Princeton Conference on Information Sciences and Systems, 148-149.

Spira (cont'd)

(1971A) "Complete linear proofs of systems of linear inequalities," IEEE Conference Record of the Twelfth Annual Symposium on Switching and Automata Theory, 202-206.

(1971B) "Complete proofs of systems of linear equalities," to appear.

(1971C) "On the number of comparisons necessary to rank an element," to appear.

Spira, P. M., and Arbib, M. A.

(1967A) "Computation time for finite groups, semigroups, and automata," IEEE Conference Record of the Eighth Annual Symposium on Switching and Automata Theory, 291-295.

Steinhaus, H.

(1958A) "Some remarks about tournaments," Calcutta Mathematical Society, Golden Jubilee Commemoration Volume, Part II, 323-327.

Sterbenz, P. H., and Fike, C. T.

(1969A) "Optimal starting approximations for Newton's method," Math. Comp., Vol. 23, 313-318.

Stockmeyer, L.

(1971A) See Paterson and Stockmeyer.

Stone, H. S.

(1971A) "Parallel processing with the perfect shuffle," IEEE Transactions on Computers, Vol. C-20, 153-161.

Strassen, V.

(1969A) "Gaussian elimination is not optimal," Numer. Math., Vol. 13, 354-356.

(1971A) See Schonhage and Strassen.

(1972/73A) "Berechnung und Programm," to appear.

Strassen (cont'd)
(1973A) "Vermeidung von Divisionen," to appear.

Tarjan, R.
(1971A) See Hopcroft and Tarjan.

(1971B) See Hopcroft and Tarjan.

(1971C) See Hopcroft and Tarjan

(1971D) "Depth-first search and linear graph algorithms," Twelfth Annual Symposium on Switching and Automata Theory, IEEE, 114-119.

(1972A) See Hopcroft and Tarjan.

(1972B) "An efficient planarity algorithm," Ph. D. Thesis. Technical Report STAN-CS-244-71, Computer Science Department, Stanford University, Stanford, California.

Todd, J.
(1955A) "Motivation for working in numerical analysis," Comm. Pure Appl. Math., Vol. 8, 97-116.

Toom, A. L.
(1963A) "The complexity of a scheme of functional elements realizing the multiplication of integers," Dokl. Akad. Nauk SSSR, Vol. 150, 496-498 (Russian). English translation in Soviet Math. Dokl., Vol. 4, 714-716.

Traub, J. F.
(1964A) Iterative Methods for the Solution of Equations, Prentice-Hall, Englewood Cliffs, New Jersey.

(1971A) "Computational complexity of iterative processes," Computer Science Department, Carnegie-Mellon University, Pittsburgh, Pennsylvania, Report CMU-CS-71-105.

Tutte, W. T.

(1966A) Connectivity in Graphs, Oxford University Press, London.

Ullman, J. D.

(1969A) See Hopcroft and Ullman.

(1971A) See Hopcroft and Ullman.

Val'skii, R. E.

(1959A) "The smallest number of multiplications necessary to raise a number to a given power," Problemy Kibernet, Vol. 2, 73-74 (Russian). English translation in Problems of Cybernetics, Vol. 2, 395-397.

Van der Waerden, B. L.

(1970A) Algebra, Vol. 1, seventh edition, translated by F. Blum and J. R. Schulenberger, Frederick Ungar Publishing Company, New York.

Van Voorhis, D. C.

(1971A) "A generalization of the divide-sort-merge strategy for sorting networks," Technical Report No. 16, Digital Systems Laboratory, Stanford University, Stanford, California.

(1972A) "An improved lower bound for sorting networks," IEEE Transactions on Computers, Vol. 21, No. 6, to appear.

Varaiya, P.

(1970A) See Cohen and Varaiya.

Varga, R. S.

(1962A) Matrix Iterative Analysis, Prentice-Hall, Inc., Englewood Cliffs, New Jersey.

Vari, T. M.

(1972A) "On the number of multiplications required to compute quadratic functions," TR 72-120, Department of Computer Science, Cornell

Vari (cont'd)
University, Ithaca, New York

Vashakmadze, T. S.
(1969A) "On some optimal algorithms," Thbilis. Sahemc. Univ. Gamoqeneb. Math. Inst. Srom., Vol. 1, 7-16 (Russian with Georgian summary).

Wakoff, G.
(1972A) See Fix, Gulati, and Wakoff.

Waksman, A.
(1970A) "On Winograd's algorithm for inner products," IEEE Trans. Computers, Vol. 19, 360-361.

Weinberg, L.
(1965A) "Plane representations and codes for planar graphs," Proceedings of the Third Annual Allerton Conference on Circuit and System Theory, 733-744.

Wells, M. B.
(1965A) "Applications of a language for computing in combinatorics," Information Processing 1965, Proceedings of IFIP Congress 65, 497-498.

Winograd, S.
(1965A) "On the time required to perform addition," J. Assoc. Comput. Mach., Vol. 12, 277-285.

(1967A) "On the time required to perform multiplication," J. Assoc. Comput. Mach., Vol. 14, 793-802.

(1967B) "On the number of multiplications required to compute certain functions," Proc. Nat. Acad. Sci. U.S.A., Vol. 58, 1840-1842.

(1968A) "How fast can computers add?" Scientific American, Vol. 219, No. 4, 93-100.

Winograd (cont'd)

(1969A) "The number of arithmetic operations required for certain computations," Proceedings of the Third Annual Princeton Conference on Information Sciences and Systems, 146-147.

(1970A) "On the number of multiplications necessary to compute certain functions," Comm. Pure Appl. Math., Vol. 23, 165-179.

(1971A) "On the multiplication of 2×2 matrices," Linear Algebra and Its Applications, Vol. 4, 381-388.

Winograd, S., and Wolfe, P.

(1971A) "Optimal iterative processes," IBM Watson Research Center, Yorktown Heights, New York, RC 3511.

Wolfe, P.

(1959A) "The secant method for simultaneous nonlinear equations," Comm. ACM, Vol. 2, 12-13.

(1971A) See Winograd and Wolfe.

Wong, E.

(1964A) "A linear search problem," SIAM Review, Vol. 6, 168-174.

Wright, E. M.

(1938A) See Hardy and Wright.

(1960A) See Hardy and Wright.

Zalesskii, A. E.

(1965A) "Reduction of certain combinatorial problems to integer linear programming," Vesci Akad. Navuk BSSR Ser. Fiz.-Mat. Navuk, No. 3, 24-28 (Russian).

SUBJECT INDEX

A

G

H

I

M

N

O

P

Q

R

S

T

NAME INDEX